Cereal Grains

Cereal Grains

EVALUATION, VALUE ADDITION AND QUALITY MANAGEMENT

D.C. Saxena
Professor
Department of Food Engineering & Technology, SLIET
Longowal–148 106
Sangrur, Punjab

C.S. Riar
Associate Professor
Department of Food Engineering & Technology, SLIET
Longowal–148 106
Sangrur, Punjab

Sukhcharn Singh
Associate Professor
Department of Food Engineering & Technology, SLIET
Longowal–148 106
Sangrur, Punjab

N. Jindal
Assistant Professor
Department of Food Engineering & Technology, SLIET
Longowal–148 106
Sangrur, Punjab

nipa

NEW INDIA PUBLISHING AGENCY

New Delhi – 110 034

NEW INDIA PUBLISHING AGENCY
101, Vikas Surya Plaza, CU Block, LSC Market
Pitam Pura, New Delhi 110 034, India
Phone: + 91 (11)27 34 17 17 Fax: + 91(11) 27 34 16 16
Email: info@nipabooks.com
Web: www.nipabooks.com

Feedback at feedbacks@nipabooks.com

ISBN: 978-93-81450-85-7

Composed and Designed by NIPA

Foreword

India has been a predominantly strong agricultural country. After attaining independence, it has been striving to develop its agricultural processing industry through successive 5-year plans for rapid economic growth and high employment generation. It is now recognized that India is world's second largest food producer.

In most western countries, people have organized themselves into consumer association to disseminate information and to safe guard their interests. Consumer goods are constantly tested at laboratories controlled by these associations and the test results are made available to the public. In India, unfortunately there is still no strong consumer association to make voice of consumer heard and nor there is any facility for comparative testing or informative labeling of consumer goods. Consumer is at the mercy of trader and manufacturer who is playing with the health, as number of food stuffs in the market are adulterated. Government records show that on an average 25-30% of edibles sold in various parts of country are adulterated. The consumers like to get maximum quantity for as low a price as possible. The sellers must meet the needs of the buyers to be able to exist. This is a vicious cycle. When the price of the food production is higher than the price, which the consumer is prepared to pay, seller is compelled to supply a food product of inferior quality, and thus adulteration occurs. To check such malpractice, PFA was passed in 1954 and came in force from June 1, 1955. It required that food be pure wholesome and honestly labeled.

India has achieved a commendable success in production of wheat. Northern states like Punjab, Haryana, Uttar Pradesh and Rajasthan are the major producers of wheat in the country. About 20 to 25 percent of wheat is received in the market and most of it is purchased by the Govt. Agencies at MSP. The MSP for wheat establishes a floor price for domestic wheat markets. At the same time, the Food Corporation of India (FCI) is required by law to procure all wheat delivered to its storage facilities at the MSP. Procurement of food grains at minimum support price ensures stability in farm prices. Quality wise Indian wheat ranks second in the world after Canada, but there is lack of establishment of quality standards. Indian amber coloured wheat competes well in the international market. It is far better than many of the varieties of USA, Australia, Argentina, and Europe. There is need to establish separate set of Indian wheat standards, proper segregation of varieties and proper cleaning and grading to boost processing and export of wheat.

Quality in the case of foods is of special significance, which directly affects human health and safety. The process of quality control starts from raw material, process design, product composition, packaging, storage, distribution, consumer preparation and feedback. With increasing awareness, it is necessary for the Indian food processing sector to become quality conscious. For introduction of quality system in the industry, the government needs to frame the standards in the form of ISO 9000 and HACCP. It, therefore, calls for sound technology and adequately trained personnel in handling food right from raw material through processing, packaging and transport to the consumption point.

The Food Corporation of India (FCI) is the nodal national agency for the procurement, storage, processing and transporting of foodgrains, and their further distribution through the Public Distribution System (PDS). The government fixes the procurement price as well as the distribution price of the foodgrains, and the FCI has to accept it. The government compensates the resulting loss through subsidy.

Today, we have Grain Standards as established by the International Organization for Standardization (ISO), which are considered as International Standards. Another source is Codex Alimentarius Commission (Codex) under the aegis of FAO (UN), which operates a Committee to formulate standards that are referred to as Codex Standards. Many countries have a National Standards Institution, which develops National Standards for their main grain crops and other commodities and issues specifications along with methods of testing and analysis. Several countries, which are in developing status, adopt or modify international standards into their national system.

The consumer wants reliable quality, freedom from adulteration, and safety even in staple food grains, which are becoming driving forces for brand emergence and their acceptance in business. Global experiences suggest that food based approaches are the most effective in addressing malnutrition and nutrient deficiency disorders. Food based approaches may include dietary diversification, improved food availability, nutrition education and food fortification. The Government of India has permitted contract farming of Basmati rice and also implemented Price Advisory for export of Basmati rice, which is expected to prevent the exporters from playing any mischief with Basmati shipments, particularly with regard to quality.

With changing times, food habits continuously change. With economic pressure and globalization food is moving out of the kitchen. It calls for mechanization and convenience without sacrificing affordability and quality. Cities are growing at 7-12% per year. With both the parents working, ready to eat foods, and variety snacks are needed to save on time, fuel end skill. With emerging trends of food market towards fast foods there are vast prospects for efficient utilization of cereals by exploring variety foods. Cereals constitute a good proportion of the product, which have gone global.

We have grossly under-developed quality assessment facilities with enforcement agencies, inappropriate personnel manning the inspection and quality monitoring system, undue delays in legal proceedings involving quality arbitration, procedure delays in clearing import and export shipments, delays in harmonizing the food standards and restrictions with those evolved by WHO-FAO Alimentarius Commission, and plethora of rules and regulations administered by multiple government agencies. There is also an increasing trend of blacklisting of Indian exporters and high percentage of rejection of Indian products by some major buyers, which is causing irreparable damage to the image of the country.

The need and timeliness of this book existed because of the emergence of a global market place for differentiated products that requires the traditional commodity-based production, handling and processing industry to adopt new practices that will assure the purity and quality of grain from seed to consumer. I am sure the all above points deserved prompt attention of one and all in this book. Keeping these objectives in view, I am happy to say that a successful attempt has been made in the form of this present book.

Dr. B.S. Bisht
Vice Chancellor
GB Pant University of Agriculture and Technology
Pantnagar

Preface

India's agriculture sector has long been held to be a success story. The Green Revolution solved the problem of insufficient food production. But now a new crisis looms over us. The crisis now is due to an interrelated and complex set of problems. The issues arise out of low productivity, low purchasing power and inadequate infrastructure.

The GNP growth rates of the last two decades have been 5.7 per cent and 5.9 percent, respectively. This clearly suggests that atleast a 6 percent growth in GNP for the next two decades is on the cards. The steep rise in demand will be due to the resultant higher per capita consumption of cereals and of livestock products. Therefore, the agriculture growth rate must accelerate considerably.

A transformation of the national economy may occur only by integrating agriculture with other sectors. A second green revolution is required which would include development of product specific crops / varieties, value addition to the secondary processed food products, post harvest technology management and introduction of efficient marketing and distribution system. Therefore, per capita production in the agricultural sector will rise, making it a commercially viable business.

The importance of crop quality management and development has had a significant impact on the production, handling, marketing and delivery of cereals and pulses. Continuing expansion of markets for value added crops (developed through genetic engineering as well as traditional breeding) requires rapid purity and quality detection technologies, and development and adoption of best quality management practices that are practically achievable, economically viable and fully accepted in the global market place.

This book contains articles contributed by experts from various parts of the country. This book shall be a valuable addition to the literature on food grains that has been designed systematically to include basis and advanced knowledge on quality management in addition to the collection of articles. This book is a useful reference for cereal technologists, scientists, chemists, and processors of grain seeds and the persons involved in the quality management of grains. The book will be highly needful for the researchers, industries and those entire academicians related to the quality of grain and further product development.

Editors

Contents

Section – III
Value Addition

Section – IV
Post Harvest Plant Operations and Quality Management

Section – V
Product Development: Status and New Strategies

Section – VI
Storage and Handling

Section – VII
Grading and Trends of Milling Industries

Contributors

Amarjeet Kaur
Department of Food Science & Technology
Punjab Agricultural University
Ludhiana
Punjab

Archana Vaishnava
Department of Genetics and Plant Breeding
G.B. Pant University of Agriculture & Technology
Pantnagar, Uttarakhand – 263 145

Alok Kumar Srivastava
Flour Milling, Baking & Confectionery Technology Department
Central Food Technological Research Institute, Mysore – 570 020
Karnataka

A.K. Bakshi
Department of Food Science & Technology
Punjab Agricultural University
Ludhiana – 141 004
Punjab

A.S. Bawa
Director, DFRL-DRDO
Mysore – 570 020
Karnataka

Amar Singh
Division of Agricultural Engineering
Indian Agricultural Research Institute
New Delhi

A. Pandeya
Division of Agricultural Engineering
Indian Agricultural Research Institute
New Delhi

Bhupendar Singh Khatkar
Department of Food Technology
Guru Jambheshwar University
Hisar – 125 001
Haryana

Bhaskar Janve
Department of Food Engineering & Technology
SLIET, Longowal – 148 106
Punjab

B.C. Srivastava
Division of Agricultural Engineering
Indian Agricultural Research Institute
New Delhi

Baljit Singh
Department of Food Science and Technology
Punjab Agricultural University
Ludhiana – 141 004
Punjab

C.S. Riar
Department of Food Engineering & Technology
SLIET, Longowal, Sangrur
Punjab

C.S. Singh
Department of Extension Education
JNKVV, Jabalpur
Madhya Pradesh

Deepa Srivastava
Biotechnology, Division of Applied Sciences
Netaji Subhas Institute of Technology
New Delhi – 110 045

D.C. Saxena
Department of Food Engineering & Technology, SLIET, Longowal, Sangrur
Punjab

D.K. Sharma
Department of Processing and Agricultural Structures
Punjab Agricultural University
Ludhiana – 141004, Punjab

D.P. Saini
Department of Genetics and Plant Breeding
G.B. Pant University of Agricultural & Technology
Pantnagar, Uttarakhand – 263 145

D.S. Sogi
Department of Food Science and Technology
Guru Nanak Dev University
Amritsar – 143 005
Punjab

G.M. Beigh
Division of Post Harvest Technology
Sher-e-Kashmir University of Agricultural Sciences and Technology
Shalimar Campus
Srinagar – 191 121, Jammu & Kashmir

Gagandeep Singh
Department of Food Science and Technology
Guru Nanak Dev University
Amritsar – 143 005, Punjab

Geeta Thakur
Department of Extension Education
JNKVV, Jabalpur
Madhya Pradesh

G.S. Padda
Department of Food Science & Technology
Punjab Agricultural University
Ludhiana – 141 004
Punjab

G.S. Mahal
Department of Food Science & Technology
Punjab Agricultural University
Ludhiana – 141 004, Punjab

H.P.S. Nagi
Department of Food Science & Technology
Punjab Agricultural University
Ludhiana – 141 004, Punjab

H. Kaur
Department of Food Science & Technology
Punjab Agricultural University
Ludhiana – 141 004, Punjab

Harsimran Singh
Department of Food Engineering & Technology
SLIET, Longowal- 148 106
Punjab

H.R. Naik
Division of Post Harvest Technology
Sher-e-Kashmir University of Agricultural Sciences and Technology
Shalimar Campus, Srinagar – 191 121
Jammu & Kashnir

Jagdish, S. Nikhade
SRF, Division of Agricultural Engineering
IARI, Pusa
New Delhi – 110 012

J.P. Jaiswal
Department of Genetics and Plant Breeding
G.B. Pant University of Agricultural & Technology
Pantnagar, Uttarakhand – 263 145

K.A. Anu Appaiah
PG Department of Microbiology
Mangalore University, Cauvery Campus
Madikeri – 571201
Karnataka

K. Prasad
Department of Food Engineering & Technology
SLIET, Longowal – 148 106, Punjab

K.S. Sekhon
Department of Food Science and Technology
Punjab Agricultural University
Ludhiana – 141 004
Punjab

Monesh, M. Pethe
Division of Agricultural Engineering
IARI, Pusa
New Delhi – 110 012

M.S. Alam
Department of Processing and Agricultural Structures
Punjab Agricultural University
Ludhiana – 141004, Punjab

Manoj Pandey
Department of Food Engineering & Technology
SLIET, Longowal- 148 106
Punjab

M. Arora
Department of Processing and Agricultural Structures
Punjab Agricultural University
Ludhiana – 141004
Punjab

N. Jindal
Department of Food Engineering & Technology
SLIET, Longowal –148 106, Punjab

P.C. Grover
Department of Processing and Agricultural Structures
Punjab Agricultural University
Ludhiana – 141004, Punjab

Poonam A. Sachdeva
Department of Food Science & Technology
PAU, Ludhiana – 141 004
Punjab

P.N. Maheshwari
Harcourt Butler Technological Institute
Kanpur – 208 002
Uttar Pradesh

Rajinder Singh
Consultant, BRS Nagar
Ludhiana
Punjab

R.K. Gupta
Central Institute of Postharvest Engineering & Technology
Ludhiana
Punjab

R.C. Verma
Department of Processing and Food Engineering
College of Technology and Engineering
Udaipur – 313 001
Rajasthan

R.B. Gautam
Division of Agricultural Engineering
Indian Agricultural Research Institute
New Delhi

R.K. Vishwakarma
Central Institute of Postharvest Engineering & Technology
Ludhiana, Punjab

R.K. Goyal
Central Institute of Post Harvest Engineering & Technology
PAU Campus, Ludhiana
Punjab

R. Kaur
Department of Food Science & Technology
Punjab Agricultural University
Ludhiana – 141 004
Punjab

Savita Sharma
Department of Food Science and Technology
Punjab Agricultural University
Ludhiana – 141 004, Punjab

Sukhcharn Singh
Department of Food Technology, SLIET
Longowal, Sangrur – 148 106
Punjab

Sohan Pal
Department of Genetics and Plant Breeding
G.B. Pant University of Agriculture & Technology
Pantnagar, Uttarakhand– 263 145

Satish Kumar
Department of Processing and Agric. Structures
College of Agricultural Engineering
Punjab Agricultural University
Ludhiana – 141 004, Punjab

Sadhna Arora
Department of Processing and Agric. Structures
College of Agricultural Engineering
Punjab Agricultural University
Ludhiana – 141 004, Punjab

Sapanpreet Kaur
Department of Processing and Agric. Structures
College of Agricultural Engineering
Punjab Agricultural University
Ludhiana – 141 004, Punjab

S.M. Ilyas
Central Institute of Post Harvest Engineering & Technology
PAU Campus, Ludhiana
Punjab

Sanjeev Uppal
Department of Processing and Agricultural Structures
Punjab Agricultural University
Ludhiana – 141 004, Punjab

S.N. Jha
Central Institute of Postharvest Engineering & Technology
Ludhiana, Punjab

Sanjay Kumar Jain
Department of Processing and Food Engineering
College of Technology and Engineering
Udaipur – 313 001, Rajasthan

S.K. Jha
Division of Agricultural Engineering
Indian Agricultural Research Institute
New Delhi

Tanuja Srivastava
SLIET
Sangrur
Punjab

Umesh Kamble
Department of Food Science & Technology
Punjab Agricultural University
Ludhiana – 141 004
Punjab

U.J.S. Prasada Rao
Department of Biochemistry and Nutrition
Central Food Technological Research Institute
Mysore – 570 013
Karnataka

V.B. Kalmegh
Division of Agricultural Engineering
IARI, Pusa
New Delhi – 110 012

V.K. Sehgal
Department of Processing and Agric. Structures
College of Agricultural Engineering
Punjab Agricultural University
Ludhiana – 141 004
Punjab

V.K. Thapar
Department of Processing and Agricultural Structures
Punjab Agricultural University
Ludhiana – 141 004
Punjab

SECTION-I

QUALITY STRATEGIES, APPROACHES AND MANAGEMENT

1

Total Quality Management and Food Safety

H.P.S. Nagi and Savita Sharma

Abstract

The definition of quality is described from a very simpler one "fitness for purpose or use" to a comprehensive "the totality of features and characteristics of a product or service that bear on its ability to satisfy stated or implied needs". The concept of total quality management in food industry is adopted recently and it is the "integrated organization approach in delighting customer (both external and internal) by meeting their expectations on a continuous basis through everyone involved with the organization working on continuous improvement in all products/processes along with proper problem solving methodology". TQM is concerned with the integration of all the efforts in organization towards quality improvement, quality development and quality maintenance to meet full customer satisfaction at economic levels. TQM enhances the work culture and employee's satisfaction through participation and involvement and consequently the image of the organization.

INTRODUCTION

The TQM culture is an approach to improve the effectiveness and flexibility of organization as a whole. It involves active participation of every employee in every department of an organization and integrates efforts to achieve functional goals. It is the recognition of employee rights and needs, redesign of work to benefit people, participative decision-making and problem solving and integrated approach to change. TQM provides organizational culture to ensure things are done right first time.

The beliefs, behaviour and values of traditional systems and TQM approach are very different. During the first half of the twentieth century, customers expected to pay extra for quality, however, in today's highly competitive business climate; quality is no longer an option. It is positive requirement without which an organization cannot survive. Quality products and services demand total a commitment and a new management system. TQM emphasized the committed customers, improved productivity, reduced costs, improved certainty in operations, improved company image, dedicated management and increased employee participation. The only way to achieve these is through commitment by management starting with the top. Now a days quality plays an important role in day-to-day business decisions. Quality performance is being widely used as an important part of personal performance appraisals.

Total quality management (TQM) is an integrated organizational approach in delighting customers by meeting their expectations on a continuous basis through everyone involved with the organization working on continuous improvement in all products/processes along with problem solving methodology. TQM is the latest to be adapted by worldwide from among the quality evolution phases. It is achieved though quality assurance department of the organization through statistical problem solving techniques.

TQM is concerned with the efforts of the organization towards quality improvement, quality development and quality maintenance to meet full customer satisfaction at lowest rates. TQM also enhances employee satisfaction through participation and involvement and consequently the image of the organization. It develops participative culture and each employee can directly involve in decision making in the area related to his work. It is organized through quality circles on voluntary basis.

Fig. 1

TQM Culture

- It is an approval to improve the effectiveness and flexibility of organization as a whole.
- It involves active participation of all employees in all departments of an organization and integrates efforts to achieve cross-functional goals as quality. This also gives satisfaction to both customers and employees.
- It involves recognition of employee rights and needs redesigning work to benefit people, decision-making and problem solving through collective approach.
- It provides organizational culture to ensure things are done right first time.

In some of the countries of the world like Japan and USA, national awards on TQM are instituted to create involvement and support of all concerned to boost the quality efforts.

Components of TQM

There are ten essential components of total quality management:

1. Quality policy and its communication
2. Team work and participation
3. Problem solving tools and techniques
4. Standardization
5. Design and implementation of quality system
6. Quality costs/measurements
7. Process control
8. Customer / Supplier integration
9. Education and training of employees
10. Quality audit and review for improvement

Traditional and TQM approach

There is almost contrary approach was followed in traditional way of tackling quality problems. There is now a sea deep change in the beliefs, behaviour and values over adopting TQM approach.

TQM approache	Traditional approach
Beliefs	
• No workers, no managers, only facilitators	• Blue-collar employees-workers White-collar employees–Managers
• Employees join voluntarily to look at and team members problems	• Workers participation must be legislated
• Involvement = Participation + commitment + pride	• Participation as it is convenient for bargaining.
• Addresses organizational needs and consumer requirements.	• Primarily welfare needs of the employees.
Behaviours	
• Integrated, cooperative style.	• Conflict win loose style
• Workers are empowered and do what is right.	• Each fight who is right
• Family oriented relationship	• Economic relationship
Values	
• Openness	• Secretive
• Trust	• Distrust
• Discipline	• Lack of discipline
• Patience	• Short sightedness
• Respect	• Hatred

TQM now

During the first half of the twentieth century, customers expected to pay extra for quality, however, in today's highly competitive business climate; quality is no longer an option. It is a positive requirement without which an organization cannot survive. Quality products and services demand total commitment and a new management system. TQM emphasized the committed customers, improved productivity, reduced costs, improved certainty in operations, improved company image, dedicated management and increased employees participation. The only way to achieve there is through commitment starting with the top. Now a days quality plays an important role in day-to-day business decisions. Quality performance is being widely used as an important part of personal performance appraisals.

Teamwork for quality

Some problems or issues are beyond the control of any one individual so some form of teamwork to tackle those problems is suggested. The use of team approach to problem solving has many advantages over allowing individuals to work separately on problems. By forming a team of concerned workers:

- A great variety of problems may be tackled, which are beyond the capability of any one individual or even one department.

- The problem is exposed to a greater diversity of knowledge, skill and experience.
- The approach is more satisfying to team members and boosts morale.
- Cross-departmental or functional boundaries can be dealt with more ease.
- The outcome of team recommendations is more likely to be implemented than individual suggestions.

Quality improvement teams

It is a group of people with a appropriate knowledge, skill and experience who are brought together by management specifically to tackle and solve a particular problem, usually on a project basis. They are cross-functional and often multidisciplinary. The quality improvement teams go a step further than "task force" to expand the traditional definition of process to include the entire production or operating system. This involves paper work, communication with other units, operating procedures and the process equipment itself.

Quality tools

- Stratify data
- Pareto diagram
- Cause and effect diagram
- Histogram
- Scatter diagram
- Check sheets
- Control charts

These have been replaced with the following new tools based on planning, goal setting and problem solving.

- Affinity diagram
- Matrix diagram
- Matrix data analysis
- Relations diagram
- Tree diagram
- Relations diagram
- Arrow diagram

How to go in for TQM

Following are the number of steps in introducing total quality management in any organization. Some of the steps may be combined or the sequence may be changed depending upon size and type of organization and its priorities.

1. Literature compilation of TQM and company's situation.
2. Interacting with outside professionals/consultants through workshops/ seminars.
3. Organization of appreciation workshop for top management and senior officials of the company.
4. Formal declaration of TQM introduction in the company by CEO.
5. Establishment of steering committee with clear mandate
6. To define quality policy.
7. Ensuring communicating policy to all concerned in the organization.
8. Organizational frame work in terms of assigning TQM responsibilities.
9. Quality awareness though trainings / workshops.
10. Forming quality improvement teams (QIT) to identify, analyze and problem solving techniques.
11. Training of QIT
12. Initiation of small group activities such as quality circles.
13. Documenting experiences of quality circles and quality improvement teams.
14. TQM workshops to identify inputs, supplier's outputs and customers internally.
15. Setting up of operational and control measures and standards.
16. Review of documentation process and quality records.
17. Identifying problem areas and weaknesses and upgrading the system.
18. Working out specific quality cost measurement system to match with existing financial management of the company.
19. Conducting quality review, follow up and quality audit.

Food Safety

In the recent part, the food safety has been given a top priority in the world all over and the food processors do their best to supply safe, high quality and vigorously inspected food to the consumers. Even then the public health problems are reported frequently especially from under developed countries.

Food borne diseases are the result of unintentional abuse of the food in the places where it is handled. People in general do not know about the delicate nature of food and how and where to store the food. Some organizations continuously work for the awareness of the common people. Some of the generic food safety messages to the consumers are listed below:

1. We all have a role in food safety and we should not expose the food where microorganisms multiply under favourable conditions.
2. Buyers must be aware while purchasing food items and must check them date of manufacturing, expiry and overall conditions of the pack.
3. Purchased food must be stored at proper temperature. Refrigeration must work at 4°C and freezers at –18°C.
4. We must clean the tools, utensils, worktops and hands frequently while preparing good.
5. When we leave any doubt regarding any food we should throw it out rather than giving to someone.
6. Thawing of foods should be in refrigeration, under running water or in microwave oven.
7. Foods must be cooked properly to doneness.
8. We must not spread the food around.
9. Hot foods are always better than normal temperature placed foods.
10. Frozen foods must be purchased at the end but at home they should be used first.

2

Wheat Protein Composition and its Relation to Quality of Indian Traditional Products

U.J.S. Prasada Rao

Abstract

Wheat is unique among cereals because of its ability to form viscoelastic dough. Variety of products such as leavened and unleavened breads, pasta, cookies, and breakfast cereals are prepared using wheat. The viscoelastic property of dough is mainly due to the presence of gluten proteins viz., gliadins and glutenins. Variations among these proteins play a significant role in determining the quality of wheat. Although considerable studies have been carried out on quantity and quality of protein and their relation to bread quality, very little work has been reported on their relation with quality of Indian traditional products like chapati and tandoori roti. Results indicate that ratio between gliadins and glutenins are important in determining the chapati quality. It has been found that varieties having 5+10 HMW glutenin subunits with IBL/ IRS chromosomes are more suitable for the preparation of chapati and roti. This article is focused on the relation between protein and its subunit composition and the quality of chapatti and roti quality.

INTRODUCTION

Wheat is unique among cereals and a wide variety of products like bread, cookies, chapati, puri, parrota, noodles, cookies, and flakes are prepared using its flour. Wheat flour derived from hexaploid *Triticum aestivum* wheat varieties, are used for the preparation of bread, chapati and other Indian traditional products.

Tetraploid *Triticum durum* wheat varieties are used for the preparation of pasta products. It is well established that the wheat proteins, especially gluten proteins play an important role in controlling the quality of various bakery products. This article is mainly focused on the relation between protein composition and roti and chapati making quality.

Protein composition

The protein content of wheat varies depending on the variety and cultivation conditions. The range of protein content in wheat is generally found to be 10-15%, but protein content as high as 21% and as low as 6.5% have also been reported (Huebner *et al* 1976, Laszity 1984).

The wheat proteins can be broadly classified into two groups. They are:

1. gluten proteins (storage proteins) and
2. non- gluten proteins (albumin, globulin).

Approximate distribution of these proteins and their physical properties are given in Fig 1.

In general, non-gluten proteins, water-soluble albumins and salt soluble globulins, constitute 10 to 22% of total flour protein and unlike gluten proteins these proteins are poor in proline and glutamic acid but rich in lysine and arginine (Bushuk, 1986). The majority of the non- gluten proteins are monomeric and their molecular weight (MW) is mostly lower than 25,000, although a significant proportion has a MW between 60,000-70,000 (Singh *et al*, 1990). Most of the non-gluten proteins are metabolic or structural proteins. Gluten proteins (storage proteins) contribute up to 80% of total flour proteins and these proteins are classified as gliadins, glutenins and insoluble glutenins. These proteins are rich in proline and glutamic acid. Gliadins and glutenins show very low solubility in water or dilute buffer solutions due to high content of non-polar amino acids and glutamine, low content of charged amino acids. Gliadins are monomeric in nature and their MW ranges between 30,000 to 80,000. However, high MW gliadin, which is polymeric in nature, has MW more than 100,000.

Glutenin is a mixture of many high molecular weight protein with molecular weight ranging from several hundred thousands to several million (Kasarda, 1989). The molecular weight of glutenin subunits range from 30,000-120,000. The subunits are classified into HMW subunits, which range from 80,000 to 120,000 and LMW subunits 30,000-70,000. More than 20 HMW glutenin subunits are detected so far in wheat varieties and newer HMW glutenin subunits are discovered frequently. More than 40 different LMW glutenin subunits were identified in different wheat varieties and these subunits are less characterized. The number of HMW glutenin subunits in a variety may vary from 3 to 6 and the number of different LMW glutenin subunits in a variety may vary from 7-16 (Gupta and Shepherd, 1990).

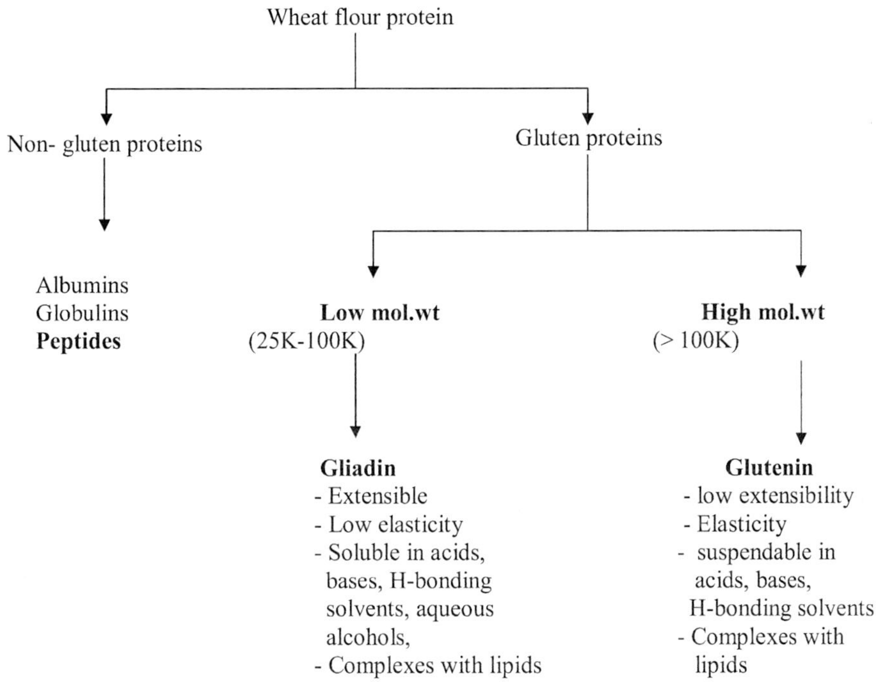

Fig. 1: A schematic representation of the protein fractions of wheat flour

The roles of individual glutenin components in dough functionality are complex. The effect of gliadins on baking quality is unclear. Some studies point out a positive effect but negative effect has also been reported (Fido *et al* 1997, Weegles et al., 1994). Fractionation and reconstitution experiments showed clearly that the glutenin proteins determined variation in bread making performance. The dough elastic strength properties are ascribed to the glutenin polymers. The large size of the glutenin polymer provides continuity to the dough system. It was reported that glutenin subunits are linked by both inter and intra molecular disulphide bonds and non-covalent interactions. These interactions are responsible for elasticity of glutenin molecule (Prasada Rao and Nigam, 1987).

Studies from different laboratories have shown correlations between presence of particular HMW glutenin subunits and baking performance of wheat flour (Branlard and Dardevet, 1985; Payne 1987; Kolster and Vereijken 1993). Genes at three loci, Glu A1, Glu B1 and Glu D1 on chromosome 1A, 1B and 1D encode these subunits respectively. It was reported that each locus exhibits extensive allelic variation. Payne (1987) reported that Glu-1 quality scores for different HMW glutenin subunits and showed good correlation between the sum of Glu-1 quality scores and bread making quality. It has also been reported that combining Glu-1 scores with protein quantity and quality characteristics improved the correlation characteristics. Recent studies using transgenic plants demonstrated that alterations of HMW glutenin subunits impacts bread making quality (Dupont and Attenback, 2003).

Correlation between protein characteristics and chapati quality

Major part of the wheat produced in India is primarily used for preparation of chapati, which form a traditional food item in the diet of majority of Indian sub-continent population. Chapati is a single layered flat bread made from unleavened dough. Good chapati must be soft, pliable and puffy and those characteristics are probably due to gluten proteins. It has been reported that quantitative and qualitative differences in proteins influenced the chapatti making qualities of different varieties (Ram and Nigam, 1981; Srivastava *et al*, 2003). Ram and Nigam (1981) evaluated the chapatti making quality of different varieties and correlated with gluten protein composition. They concluded that the relative quantities of gliadins, glutenins and insoluble glutenins (residue protein) play an important role in determining chapati quality. Wheat varieties having about equal quantities of these proteins yield better chapattis than varieties in which excessive amounts of one or the other gluten components are present. Wheat varieties having more of residue protein give leathery chapattis and varieties having more of gliadins give relatively stiff chapatis.

Srivastava *et al* (2003) studied various Indian wheat varieties for their chapati making quality and correlated it to protein characteristics. It was reported that the puffed height, which is an important parameter for chapatti, was positively correlated with protein content and gluten content. Cultivars having 5+10 HMW glutenin subunits at the Glu-1D chromosome, with IBL/IRS rye chromosome with a Glu-1 score of 6 yielded excellent chapati. In general, varieties with 2+12 HMW glutenin subunits yielded poor chapati quality.

Correlation between protein characteristics and tandoori roti quality

About 15-20 % of the wheat produced in India is used for production of tandoori roti. Apart from India, it is consumed in many Asian countries. It is prepared using whole-wheat flour. Tandoori roti is one of the traditional bakery products consumed in India. This is a single layered flat bread made from whole wheat flour. It is prepared by baking the sheeted dough in a special type of oven known as ***tandoor*** for about 37 s at 425°C.

Saxena *et al.* (1997) studied the correlation between the gluten proteins and tandoori roti quality from different Indian wheat varieties. Varieties having total protein content of about 10% with high gluten content and about 5% total glutenin contents was found to yield a good roti quality. It was further reported that gliadin content has negatively correlated with the roti quality. Studies on the correlation between the HMW- glutenin subunit composition and roti quality indicated that varieties having 5 + 10 subunit at the Glu-1D chromosome with a Glu-1 score of 6 yielded better roti quality. Presence of IBL/ IRS rye translocation with 5+ 10 HMW glutenin subunits yielded a good roti quality. Based on the chemical, physical and rheological studies, it was suggested that medium hard wheat varieties are more suitable for the preparation of tandoori roti.

The studies presented here clearly indicate that protein content and quality of HMW glutenin subunit composition has influence on roti and chapatti making quality. It can also be noted from these reports although most of the varieties having 5+12 HMW glutenins subunits gave good chapatti / roti quality few varieties having the similar subunits gave poor quality also. It indicates it is necessary to look some other parameters like concentration of HMW subunits. LMW sub units composition was given less attention with regard to baking properties. Quantitative and qualitative differences in LMW glutenin subunits and the difference in ratio between HMW or LMW glutenin subunits also need to be looked into.

REFERENCES

Branlard, G. and Dardevet, M. (1985). J Cereal Sci 3, 345.
Bushuk, W. (1986). Cereal Foods World 31, 218.
Dupont, F.M. and Attenbach, S.B. (2003). J Cereal Sci 38, 133.
Fido, R.J., Bekes, F., Gras, P.W. and Tatham, A.S. (1977). J Cereal Sci 26, 271.
Gupta, R.B. and Shepherd, K.W. (1990). Theor Appl Genet 80, 65.
Huebner, F.R and Wall, J.S. (1976). Cereal Chemistry 53, 258.
Kasarda, D.D. (1989). Glutenin structure in relation to wheat quality. In: Wheat is unique. St. Paul, MN, USA, AACC Inc. p.277.
Kolster, P. and Vereijken, J.M. (1993). Cereal Foods World 38, 76.
Lasztity, R (1984). The importance and general characterization of cereal proteins. In: The Chemistry of Cereal Proteins. CRC Press, Inc. Boca Raton, Florida. PP 3.
Payene, P.I. (1987). Plant Physiol 38, 141.
Prasada Rao, U.J.S. and Nigam, S.N. (1987). Cereal Chem 64, 168.
Ram, B.P. and Nigam, S.N. (1981). J Food Science 47, 231.
Saxena, D.C, Prasada Rao, U.J.S and Haridas Rao, P. (1997). J Sci Food Agric 74, 265.
Singh, N.K, Donovan, G.R, Batey, I.L and Mac Ritchie, F. (1990).Cereal Chem 67, 150.
Srivastava, A.K., Prasada Rao, U.J.S. and Haridas Rao, P. (2003). J Sci Food Agric 85, 225.
Weegles, P.L., Marseille, J.P., Bosveld, P. and Hamer, R.J. (1994). J Cereal Sci 20, 253.

3

Development of Testing Protocol for Processing Machinery to Enhance Grain Quality

R.K. Gupta and R.K. Vishwakarma

Abstract

There are enormous food processing industries comprising of grains and oilseeds milling, fruits and vegetables processing, spice grinding and other food processing units in India. Also, there are many seed processing companies in the country, which are engaged in the processing and production of good quality seeds of different crops besides National Seeds Corporation and State Seed Corporations. All industries are using different type of machinery and equipment like cleaner, grader, expeller, milling machines, polishers, dehusking machines, size reduction equipment, different type of grinders, separators, treatment machinery, mixers, pulverizes etc. Besides this equipment, material-handling equipment like conveyor (belt, chain or screw type), elevators, holding bins and other handling equipment are also frequently used in the processing plants. To meet the above requirement many firms are also involved in the manufacturing of these equipment and machinery. Since, most of the seed processing equipment, dhal milling plants, material handling equipment are manufactured by the unorganized sector and hence equipment and machinery fabricated by them are not of very good quality. However, Bureau of Indian Standards has formulated the codes of manufacturing for these equipment and machinery. Seed processing equipment manufactured in India are generally copy of imported processing equipment. The manufacturers do not use materials of good quality. Framework of the machines is very poor. These machines are not also tested for their efficiency, workability strength, and other parameters. Most of the time performance of these machines is not up to the mark and wide variations in the performance is observed. In fact, these machines should be tested for their structural strength, capacity specified by manufacturer, efficiency, type of material to be used as per BIS specifications, workability, ease of operation, safety, ergonomics, level of noise produced by the machine, operating conditions, type of standard accessories used, alignment of the moving parts, balancing of the moving parts, total weight of the machine, repair facilities, etc. As for as testing of material quality and manufacturing of the equipment/machine is concern; industries may be having their own testing facility. But other parameters like efficiency, ease of operation, safety, ergonomics, level of noise produced by the machine, operating conditions etc to be tested for better performance of the machine and equipment in the field. More specifically, there is a need to develop testing procedures and protocol for testing of grain processing machines such as cleaner cum grader, indented cylinder grader, gravity separator, destoner, etc. including material handling equipment such as bucket elevator and belt conveyor for their operating performance. Besides, dust collector, installed in the grain processing mills including dhal mills should also be tested for their efficacy. The codes for the acceptable level of noise pollution should also be evolved. This will lead to production of good quality grains for use such as and its further processing. The issues like ease of operation, safety, ergonomics, level of noise produced by the machine, operating conditions, etc. for better performance of the machine and equipment have been discussed in the paper. This paper not only focus the issues related to development of testing protocol for processing equipment and machinery for production of good quality grain but also provides information which may help in modifying the equipment and machinery to suit the requirement of the food related industries.

INTRODUCTION

Grains are among the major commodities for feeding mankind. The cleaning, drying, and storage of grains are post harvest operations required to maintain their product quality. The commercial value of cereals dependent on how suitable the grain is for use in consumer products. This in turn is dependent on the functional and nutritional quality of the grain. Cereal grains are subject to quality loss during primary processing, storage, and transportation often resulting in considerable diminution in grade and value. Quality loss is of concern to many persons in the grain industry including produces, elevators managers, shippers, regulators, exporters, and purchasers both domestic and foreign. Almost all the food, feed, fibre and fuel commodities go through a number of post harvest operations such as cleaning, grading, separation, drying, storage, milling, food engineering operations, packaging, transport and marketing before reaching to the consumers. The importance of post harvest technology and associated equipments is increasing and will continue to increase as agricultural productivity increases. To meet the standard of good quality grains for further refining, testing protocol for processing equipment and machinery should be evolved as a National Facility so that perfection in processing equipments and machines could be achieved for production of good quality grains.

NEED OF TESTING

At present, many firms manufacture seed processing equipments in India. The machines manufactured in India are not designed by the manufactures. These are copied from the imported machines as per their requirement. The manufactures have no system to select the material for construction of different parts of the machine. These machines do not give the quoted efficiency. Again, the different items used in the machine like bearings, brackets, belts, belt drives, pulley, motor, drives, nuts & bolts, etc. are not used as per requirement. The welding of joints are not proper. Moreover, continuous welding is usually not done. Finishing of the joints, etc. are not proper. Location of different parts of the machine is not as per the requirement but these are fitted on the basis of ease of manufacturer.

Fans and blowers used by the manufactures are made by them and are not statically and dynamically balanced. Wood and wooden frames used by the manufactures are not selected properly. The manufacturers do not provide any information on selection of sieve for different crops and varieties. Sieves fitted with machines usually sag after some time which reduces cleaning efficiency and cause clogging of screens also. Brushes or balls used for screen cleaning are not selected properly. Screen knockers create lot of noise pollution. Feeding mechanisms are not designed as per the requirement of the crops. This leads to improper feeding of raw material on the screen and cleaning efficiency is reduced. Leakage of grains is observed from different points and framework of the machines are not strong enough to withstand the entire load and hence vibration in machine components takes place resulting noise pollution. Also, this reduces the useful life of the machine. The machine parts, which has to be replaced regularly like screens, etc. are not fitted in a fashion that they can be changed easily.

Some bearings, bushes are fitted at a place where the oiling and greasing is very difficult resulting problem in repair and maintenance. Though, the manufactures provide belt guards for safety, but some of the moving parts are not fitted with guards and may cause accidents. Control mechanism provided with the machines often does not work and hence, uniformity in feeding does not took place and efficiency of the machine is reduced. Thus, if the machines should tested for strength, materials selection, type and quality of material used, ergonomic point of view, ease of operation and maintenance, proper design, safety devices, operating principals, strength of materials and moving parts, etc. the performance of the machine can be improved. This will also help in improving the machine efficiency. The machines can be operated easily and will be free from any kind of pollutions. The operator will feel safe during the operation. The life of the machine will be more with easy repair and maintenance. This will also lead to least mechanical damage to the seed during the processing and minimal loss of valuable seed material during processing will occur. Thus the quality of primary processed grains will improve. The machines should also be tested for their suitability for different grains.

TESTING PARAMETERS

Before testing any seed processing equipment and machinery, there is a need to set the parameters on which, the machine will be tested. Each equipment and machine used in the seed processing has different type of parts. Hence, some test parameters should be evolved which may be tested as performance indices for different machines. More, specifically, there is a need to prepare testing procedures for each equipment and machine used for processing of grains. In general the parameters which has to be tested for judging the performance of the processing machines such as cleaner cum grader, indented cylinder grader, gravity separator, destoner, etc. are described below;

General description of the machine

A report should be prepared in which the general information of the machines are given. It may contain information on overall dimension of the machine, type of fan/blowers used, machine power, drive mechanisms, type of feed hopper, principle of processing, number of screens, fans etc. used, Operating principals should also be described in the in the report or manual. It should also contain information on efficiency machine, capacity, grain for which it can be used and other relevant information.

Feeding unit and mechanism

Feeding unit of the machine is important as it decides the capacity of the machine and should be tested for its design.. It is also important that feeder is placing the material at which part of the machine. It should also be checked whether the feeding unit is feeding the material uniformly or not. Because uniform feeding to the screens in grader improves its efficiency and quality out-put. It should also be checked whether any damage to the grain is taking place in the feeding section itself or not. . It should also be checked for its slopes and height from the grading sieves, etc. Because the height of out let of feeding unit of the machine from screen is linked with the impact on the grain.

Type of material used

Parts of seed processing equipment are made of different type of materials like mild steel, stainless steel, cast iron, carbon steel, wood, nylon brushes, rubber balls, etc. Test procedures must include the information on requirement of material for each part should clearly defined and same material should be used for manufacturing that part of the equipments and machines. Wood should also to be checked and insured that it would be free from any defects and type of wood to be used should also be defined. Each material should be tested for its strength, composition, durability, etc.

Structural Strength

The machine structure should be strong enough to with stand all the loads and shocks coming to the frame. It should not vibrate during operation in long runs. The joints of the frame should be welded properly. Spot welding should not be permitted in the frame. Joints should be made with proper sealing material. At present manufactures are using synthetic rubber (used) for sealing. This should be replaced with proper sealing material and its testing should be insured. The structure should be stable and balanced. Procedure for testing the structural strength should be prepared. The test procedures need to be prepared taking into consideration of all above characteristics.

Balancing of different parts

Balancing of fans and blowers is essential since unbalanced fans and blowers do not provide uniform airflow rate and enormous load is exerted on the bearing and motors which may result in failure of these parts. Every moving part of the machine should be balanced properly. Fans should be balanced statically and dynamically. These parts should be tested and test code should be prepared. Strict testing of these parts is essential as poor performance of the blowers affects the performance of cleaning machines severely.

Components of the machine

Each seed processing equipment and machine consists of number of components. Like feed hopper and feeding mechanism, bearings, brackets, cam and cam drives, bushes etc. Each component should be checked separately and test procedures should be prepared. Every bearing, brackets, bushes and other components should be of approved good quality. Fittings of these components should also be checked.

Control mechanisms

In the processing machines, there is always a provision for changing the speed, oscillation frequency, stroke length, speed of fans, inlet, and outlet openings, openings for air etc. These control mechanisms should be checked for their working whether they are working as per the specifications or not. It should also be cheeked during full load processing. The test procedure should be prepared in such a way that it provides information of no load as well as under full load conditions.

Screens shoe and screen cleaning mechanism

Screens are the most important component of the seed processing equipment and machines. Effectiveness of screens plays very important role in overall efficiency of the machine. Capacity of the machine depends upon the slope of the screen as well as oscillating speed of the screen. The less inclination of the screen may cause over loading whereas more inclination of screen may yield enhanced capacity but effectiveness of screen may decreased. Hence, screen should be checked for its strength, screen material, uniformity of openings, length and width, slope etc. It should also be checked whether screens are sagging during operation of the machine. The ease of replacement of screens should also be checked and reported. It should also be checked whether the screen assembly (shoe) oscillation is uniform or not.

Testing of screen cleaning mechanism is essential. The balls, brushes, or screen knockers used for cleaning the screens must be tested for their material and operation. It is also important to check whether screen is fitting in the grooves properly or not. There should not bee any gap after inserting the screen otherwise vibration of screen may occur and grading efficiency may also be affected. The material used in brushes/balls etc. should be of kind that they do not damage the shape of the Screen. It should also be checked whether shape of the screen has been changed or not during operation.

Operator's Manual

The operator's manual provided by the manufacture should be cross-examined for the information's contained in it. This should contain the general information of machine, function of machines, adjustments, and settings for various crops, repair and maintenance, service, safety, types of screens for various crops and varieties etc. in detail. All the information also is described with suitable flow chat and diagrams.

Operator's Safety

The machine safety guards provided by the manufacture should be tested and checked. It is also important to provide information about safety norms. These safety guards should not hamper the operation of the machine and creates obstruction to the operator. The wirings, starters etc. should not be fitted on machine to avoid electric shocks while in operation.

Mechanical History

The mechanical history of the machine should be observed during testing if any part of the machine is creating problem during operation. Tension of belts, vibrations, shocks, leakages of grain from any part of the machine, sealing, cracks developed in any part of the machine should be checked and reported.

A detailed evaluation report of the machine on the points mentioned above should be prepared. It is also recommended that the manufacture should provide an operator's manual containing information on operation, adjustment, safety, lubrication, maintenance, and warranty. Providing standard equipment and safety deals to alert operator of potentially hazardous areas to ensure the smooth functioning of the processing units.

Quality of work

The processing equipment and machines should be tested for their performance for all the grains for which manufacturers specifies the machine. It is important to check whether the machine is removing contaminants to specific level or not. Level of impurity, type of possible contaminants, flow rate, etc. should also be standardized and mentioned for each grain. Specified capacity should also be tested. It is also important to generate data on machine efficiency at different level of airflow rates, feed rate, and other adjustments. Efficiency and effectiveness of each screen should be tested separately. The machine should also be checked whether any mechanical damage to the grain is taking place during primary processing. The unit, in which this damage is taking place, should be reported to the manufacturer. Processed seed coming out of the machine should be tested as per the rules of seed certification. Effectiveness of blowers and aspiration is important and should be tested carefully; capacity, efficiency, and other seed parameters should also be reported for better performance of the system.

Rate of work

Capacity of any seed processing equipment is dependent on type of crop contaminants and grades desired. All these points should be tested and reported.

Ease of operation and adjustment

Each machine should be tested for its ease of operation and adjustment. It is important that changing screens, cylinders or other items should not be time consuming and awkward. The machine should not be stopped for long time during adjustments, and changes.

Power requirement

It is very important to check the power requirement of the machine as it is directly linked with energy consumption and economics of processing. It is essential to check whether motors and drives fitted are compatible or not. It should also be tested that losses in power transmission, couplings drives are how much and optimum solutions may also be suggested to the manufacturers. All the data regarding number of motors, location of motor, power etc, should be reported and specified. In other words, effective plant-layout is to be worked out.

RECOMMENDATIONS

Thus, detail evaluation of the processing equipment and machines will help a great in improving the performance of grain processing machineries. It will help in further improvement in the machine. A testing protocol should be developed for the grain processing equipment and machines for production of good quality grain and also providing information to the manufacturers for modifying the processing equipment and machines. With the development of such facility good quality efficient grain processing machines could be made available in the country to suit the requirement of the food related industries.

REFERENCES

Bolz, HA and GE Hgemann (Eds). 1986. Material Handling Handbook. The Ronald Press Company, New York.

CGGR Handbook of Agricultural Engineering. Agro-Processing Engineering (Volume IV). Edited by CIGR, Published by ASAE, St. Joseph MI49085-9659 USA.

Evaluation report on Harvey Gjesdal Five in one Rotary Seed Cleaner manufactured by Simon –Day Ltd., Winnipeg, Manitoba.

Hall, CW and DC Davis, 1979. Seed and Grain Cleaning and Grading Equipment. In Processing Equipment for Agricultural Products (Second Edition), AVI Publishing Co. Inc. Westport, Connecticut.

Henderson, SM and RL Perry, 1976. Agricultural Process Engineering, AVI Publishing Co. Inc. Westport, Connecticut.

Henry RJ and PS Kettlewell.1996. Cereal Grain Quality. Published by Chapman & Hall, London.

4

Mycotoxins in Cereal Grains

Deepa Srivastava, Tanuja Srivastava, C.S. Riar, Sukhcharn Singh and D.C. Saxena

Abstract

Mycotoxins are naturally occurring chemicals produced by fungi growing on grain, feed or food. The fungi that produce mycotoxins in grains commonly include Fusarium, Aspergillus and Penicillium species. The toxins can accumulate in maturing corns, cereals, soybeans, sorghum, peanuts and other food and feed crops in the field and in grain during transportation. The toxins may occur in storage under conditions favorable for the growth of the toxin producing fungi. Diseases in animals and human beings resulting from the consumption of mycotoxins are called Mycotoxicoses. These fungal metabolites are detrimental to the health of both animals and human. Toxicity ranges from acute death to chronic diseases, cancer and reproductive malfunction. Mycotoxin contamination of food and feed grains is a serious economic problem for grain producers and processors because grain contamination impacts market.

INTRODUCTION

Mycotoxins are secondary fungal metabolites. They are chemically diverse and occur in a wide variety of substrates including foods and feeds. They may cause illness and death of man and animals if food or feed containing them is eaten. They may cause economic losses in livestock through disease and reduced efficiency of production. Many of the molds that produce the toxins are also frequent contaminants of food and agricultural products. These include members of the genera Aspergillus, Penicillium, Fusarium, Alternaria, Claviceps, Phoma, Diplodia, Cladosporium, Rhizopus, Sclerotinia, and Phomopsis.

Formation of mycotoxin is closely related to mold growth. Without mold growth, mycotoxin production will not occur. The molds can grow on crops in the field, in the grains at harvest, during transport, in storage and during processing of feeds when conditions favorable for their growth prevail. Several factors influence mycotoxin production by molds, including substrate, moisture, temperature, pH, and stresses such as drought and associated growth of other molds or microbes. For example, aflatoxin-producing *Aspergillus flavus* produces aflatoxin in higher amounts in substrate with certain amino acids, fatty acids and zinc, at 85% relative humidity (over 17% mc in corn), and at 25-27°C. Usually the toxin is not formed when other molds are present,

Fusarial mycotoxins are usually produced by the Fusarium while the crops are still in the field. Different species of Fusarium infect cereal grain causing a variety of diseases such as scab in wheat and barley and cob rots in corn. Mycotoxins may be are produced as a result of this infection. These toxins if produced remain until the grains are stored, processed and used as food or feed.

Aflatoxin production by *A. flavus* can also occur in the field. Stress due to drought and insect damage can increase the chance of the mold to invade the kernels, this invasion may be accompanied by the formation of aflatoxin.

The presence of molds in or on the foods or feeds does not automatically mean the presence of mycotoxins but, rather, that a potential for mycotoxin contamination exists. On the other hand, the absence of these molds does not guarantee that the commodity is free of mycotoxins because toxins can persist long after the mold has lost viability.

In India, grain molds occur every year to varying degrees on field corn, seed corn, white corn, and other specialty corn hybrids. Most grain mold pathogens become associated with the kernel in the field; however, under certain conditions of temperature, relative humidity, and grain moisture, these molds can grow within the colonized kernel and even spread to adjacent kernels during transport and storage. Several issues are associated with grain molds in corn including lowered grain quality, effects on human health, and effects on animal health and reproduction. Most grain mold pathogens have the potential to lower grain quality by affecting feed efficiency or grain processing characteristics. Human health concerns include allergenicity and hypersensitivity associated with the inhalation of mold spores in grain storage facilities and acute or chronic disease associated with infection by some species of grain mold fungi. In addition, some grain mold pathogens produce compounds (mycotoxins) that can be toxic to farm animals, wildlife, or humans. The presence of mold, however, does not indicate contamination of the grain with mycotoxins. Only certain strains of certain fungal (mold) species have the potential to produce these potentially harmful compounds. Grain mold pathogens are also capable of causing disease in other parts of the plant. *Fusarium graminearum* and *Fusarium verticillioides* (formerly named *F. moniliforme*) cause stalk rots, capable of producing different classes of mycotoxins within colonized kernels. Recent evidence suggests that the mycotoxins are produced during infection of the plant and although not essential, may play a role in disease development.

Common Grain Molds

Grain molds are classified into field molds and storage molds based on the percent relative humidity required for their growth. Field molds require equilibrium relative humidity above 90% (or minimal seed moisture content of 22-23%). These molds invade the seeds (generally limited to the pericarp) before harvest while the crop is still in the field, or after the crops have been cut and stacked. They may survive in stored grains but can not grow as long as the grain moisture content is below 20%. Major genera of field molds include Alternaria, Fusarium, Cladosporium, Curvularia and Helminthosporium.

Storage molds grow in grains with moisture content in equilibrium with 70-90% relative humidity (minimum moisture content of 13-13.5% in grains). These are the molds that cause mustiness, heating, kernel damage and various deterioration changes in stored grains and grain products. Several species of Aspergillus and Penicillium comprised mainly the storage molds.

Molds are fungi that grow by producing long filaments called hyphae. In general, hyphae are important to the survival and dispersal of fungi. Hyphal growth allows the fungus to colonize a food source (e.g., a corn kernel) as well as to grow from one food source to another; e.g., from root to root through soil or from one kernel to an adjacent kernel in a pile of stored grain. A network of hyphae is referred to as mycelium. This hyphal network is responsible for "cementing" kernels together in grain piles resulting in columns of grain that cannot be separated. Grain mold fungi also produce spores (conidia) capable of aerial dispersal in the field as well as within a grain storage bin . It is usually masses of spores that give the mold a characteristic color. Spores are dispersed passively by wind and rain. Insects can serve as vectors of these fungi usually by transporting the spores on the surface of their bodies; this is particularly important within grain storage bins. Managing grain storage insects can reduce contamination by grain molds and mycotoxins. Most species of grain mold fungi are well adapted to the conditions of grain production and postharvest handling and storage. They can survive long periods in storage facilities making sanitation of the facility an important part of a grain mold management plan.

The most striking external symptom of grain mold is the presence of the mold itself. The degree of growth on the kernels and the appearance of the mold (e.g., color and density) varies with the species of mold, the quality of the grain being colonized, and the prevailing environmental conditions. Incidence (the proportion of ears with mold) and severity (the proportion of infected kernels on an ear) of disease depends on many factors.

Aspergillus species tend to be more prevalent when there is drought during the latter half of the growing season. *Fusarium verticillioides* is associated with a high proportion of corn kernels under most growing conditions but *Fusarium* ear and grain mold develops more often when cool wet weather during silking is followed by hot dry weather. *Gibberella* grain mold is more prevalent in hybrids with tight husks. Unlike *F. verticillioides*, *F. graminearum* is rarely seed-borne.

Many species of fungi cause molding in grain. Most grain mold fungi are relatively non-specific; they colonize several species of plants and utilize different types of organic matter as an energy source. Most of these fungi become associated with the grain in the field although they may only become visible during storage, and then only if environmental conditions are favorable (e.g., moderate temperature and high relative humidity). *Fusarium*, *Gibberella*, *Diplodia*, *Penicillium*, *Aspergillus*, and *Cladosporium* are among the prevalent grain mold pathogens in India. These species are widespread across the Corn Belt and can be found throughout India every year. They all survive in field residue. *Fusarium spp.* and *Diplodia* are the same pathogens that cause stalk rot diseases during the growing season.

Fusarium Grain Mold

Fusarium grain mold is probably the most common grain mold pathogen in India. It is caused by three seed-borne species of *Fusarium*: *F. verticillioides*, *F. proliferatum*, and *F. subglutinans*. The role of seed-borne inoculum to the development of grain mold is uncertain. Disease symptoms and severity vary with hybrid genetics and environmental conditions. Moldy kernels may be clumped at the tip of the ear or randomly dispersed across the entire ear. Infected kernels may be pink or show a white starburst pattern radiating from the top of the kernel.

Gibberella Grain Mold

Gibberella grain mold is also common in India. It is caused by the fungus *Gibberella zeae*; the asexual stage of the pathogen is *Fusarium graminearum*. Unlike *F. verticillioides*, *F. graminearum* is rarely seed-borne. Disease symptoms and severity vary with hybrid genetics and environmental conditions. Infected kernels are usually clumped at the tip of the ear and the mold is reddish in color.

Diplodia Grain Mold

Diplodia grain mold is common throughout the Corn Belt, usually occurring weeks after silking. It is caused by the pathogen *Stenocarpella maydis* (Diplodia maydis) which can be seed-borne. Like the *Fusarium* spp. that cause grain mold, Diplodia survives on corn residue. Consequently, the incidence of Diplodia has increased in recent years as a result of the increased practice of reduced tillage. Infected kernels may show no external symptoms or be brown with small brown-to-black reproductive structures at the base of the kernels.

Penicillium Grain Mold

Penicillium grain mold is probably the second most common grain mold pathogen in India. It may be caused by several species of *Penicillium*, including *P. oxalicum* and *P. chrysogenum*. *Penicillium* species are well adapted to survival in many types of storage facility. Kernel infection can occur in the field or in storage. Symptoms range from external mold development to internal discoloration ("Blue Eye") of the embryo. Symptoms caused by *Penicillium* are easy to confuse with those caused by *Aspergillus glaucus*.

Aspergillus Grain Mold

Aspergillus grain mold is probably the least common grain mold pathogen. At least three species of *Aspergillus* can cause grain mold in corn, including *A. flavus*, *A. parasiticus*, and *A. glaucus*. Like *Penicillium* species, *Aspergillus* species are very well adapted to survival in many types of storage facility. Infection of kernels can occur in the field or in storage. Also like *Penicillium*, symptoms range from mold development on the surface of kernels to internal discoloration ("Blue Eye") of the embryo.

Mycotoxins

Most fungi produce a class of chemical compounds called secondary metabolites. These compounds have a wide range of biological activities including antibiotic (antibacterial and antifungal), acute and chronic toxicities (plant, animal, and humans), and hormone and growth regulation (plants and animals). There is much speculation but little direct evidence for a specific role of secondary metabolites in the growth and reproduction of the organisms that produce them. Most mycotoxins are secondary metabolites. Several species of fungi that colonize grains produce mycotoxins, however not all isolates of a toxigenic species produce mycotoxins and isolates capable of producing a mycotoxin do not always synthesize the toxin. Clearly, synthesis and excretion of the toxin is not required for reproduction and survival of the organism. There are several chemical classes of mycotoxins, each with a unique chemistry. Generally, the chemical class of mycotoxin is specific to a genus of fungus; i.e., usually only one genus of fungus produces a given class of mycotoxin. Only two species of *Aspergillus* produce aflatoxins and only a few species of *Fusarium* produce the class of mycotoxins known as fumonisins; however, species of two genera, *Penicillium* and *Aspergillus*, produce the mycotoxin ochratoxin.

Mycotoxins have a wide range of activities and target species (Table I). Some mycotoxins have toxic activity in both animals and plants, while others are only toxic to one or a few species. Mycotoxins vary greatly in their toxic effects and their potency. Some mycotoxins are somewhat target specific affecting primarily one organ (e.g., liver, kidney). Others have broad spectrum activity affecting several organs and tissues. For some mycotoxins, a concentration of many parts-per-million is required to elicit a toxic effect, while for other mycotoxins a concentration as low as 1 part-per-billion will elicit a toxic effect. Most mycotoxins can act alone. Some also act synergistically with other toxins causing effects greater than either toxin acting alone at concentrations less than required for either toxin alone. This can make field diagnosis of mycotoxin contamination very difficult. It is not unusual for more than one species of grain mold pathogen to be present within a single storage bin and many grain mold pathogens can be active under the same environmental conditions providing for the opportunity to have more than one mycotoxin produced in a lot of contaminated grain.

The nature of the toxic effects caused by mycotoxins on animals or human varies greatly Table 2. Some mycotoxins cause acute toxicities (i.e., immediate effect) where a certain organ (e.g., liver, kidney) loses complete or partial function; other mycotoxins cause chronic toxicities (i.e., long-term) resulting in symptoms such as weight loss and reproductive dysfunction. Still other mycotoxins impair the immune system predisposing the affected animal to a variety of infections or other ailments. For some mycotoxins damage is not permanent and affected animals can recover from ingestion if the contaminated feed is removed from the diet.

All handlers of stored grain should exercise caution when in a storage facility especially if a mold problem exists. In addition to the mycotoxin concerns described above, grain mold fungi can affect human health in several ways. Exposure to the spores of the grain mold fungi can lead to acute allergic responses (e.g., hypersensitive response leading to pneumonia) or chronic allergic responses (e.g., sinusitis, acquired sensitivity). Some species can act as opportunistic pathogens causing serious disease in people with impaired immune function (e.g., people on broad spectrum antibiotics, corticosteroids, or immune suppressants following a transplant). All personnel should wear dust masks to minimize exposure to grain mold fungi.

Table 1: Toxigenic fungi, their toxins and effected grain

Toxin	Fungus	Grains commonly Affected
Vomitoxin	*Fusarium granarium*	wheat, barley, oats, corn, rye
Zearalenone	*Fusarium granarium*	wheat, corn, barley, oats, sorghum
Fumonisin	*F.moniliforme, F.proliferatum, F. nyagmei*	corn
Aflatoxin	*Aspergillusflavus A.parasiticus*	corn, groundnuts, cottonseed, nuts
Ochratoxin	*A.ochraceus,Penicillium viridicatum*	corn, wheat, barley, oats

Table 2: Toxigenic fungi, their metabolites and target effects

Grain Mold Fungus	Toxin Produced	Toxic Effects	Species Affected
Aspergillus flavus	aflatoxin	acute toxicity (liver)liver cancerimmune suppression	many humans, animals
Aspergillus alutaceus	ochratoxin	acute toxicity (kidney) cancer	swine, poultry, human
Fusarium verticillioides	fumonisin	blind staggers pulmonary oedemaoesophageal cancer	horses wine human
Fusarium graminearum	trichothecenes	acute toxicityimmune suppression	many (not ruminants) many
	vomitoxin	acute toxicity	many
	zearalenone	reproductive dysfunction	swine
Penicillium spp.	ochratoxin	acute toxicity (kidney) cancer	swine, poultry human

Grain Mold and Mycotoxin Management

Pre-harvest

Hybrids that are less susceptible to grain mold should be planted where available in areas with perennials having grain mold and/or mycotoxin problems. Identifying the prevalent grain mold pathogens will be necessary prior to selecting the appropriate hybrids. After planting, the overall strategy for grain mold management is to minimize plant stress during and after silking. Common stresses associated with grain mold include: high moisture in mid-to-late season after a dry early season, moisture stress early in season and during grain fill, high leaf disease pressure, and insect damage.

Postharvest

All grain storage facilities should be monitored regularly to detect grain mold development. Grain molds rarely develop uniformly throughout a storage unit; development of hot spots is common. Monitoring requires a systematic sampling plan to account for the unique design characteristics of each storage facility.

Ensure Proper Storage Conditions

To minimize the potential for mold growth and mycotoxin production, grain moisture content should be reduced to less than 15 percent within 48 hours after the grain is harvested. This may be difficult to achieve in many storage facilities, especially larger capacity elevators. A storage temperature of less than 40° F should be maintained. Good air circulation throughout the storage bin is important. Storage conditions optimal for maintaining grain quality will minimize mold development and mycotoxin contamination.

Minimize Mechanical Damage

Harvest and postharvest grain handling should be designed to minimize mechanical damage. Although most mold pathogens can directly penetrate plant tissues, mechanical damage provides additional entry sites, facilitating infection and spread from kernel to kernel during shipping and storage.

Minimize Insect Damage

Like mechanical damage, insect feeding provides additional entry sites for mold pathogens, facilitating infection and spread from kernel to kernel during shipping and storage. Some insects also act as vectors of grain mold pathogens. An effective IPM-based insect management plan during the growing season as well as during postharvest shipping and storage of the grain can reduce mold and mycotoxin contamination.

Sanitation

The grain mold pathogens are well adapted to survive for long periods within grain storage facilities. Storage bins should be cleaned thoroughly each season. The type and configuration of the storage facility will dictate the nature of the sanitation required. The objective is to remove old grain and residue that could provide a harborage for these fungi.

Chemical Management

Fungicides are used to manage grain molds or mycotoxin contamination. There are many fungicidal fumigants, which are used in grain storage facilities. Formulations of propionic acid are used in some storage facilities with varied success in reducing grain mold and mycotoxin contamination.

Assay Moldy Grain

Prior to use, moldy grain should be analyzed for possible mycotoxin contamination. Depending on the intended market (e.g., food or feed, domestic or export) and the mycotoxin present, contaminated grain should be handled according to the specific recommendations for that toxin and the level of contamination. This may require segregating the grain for allowable uses. With some mycotoxins, blending with non-contaminated grain is permitted. In a few cases, the contaminated grain must be destroyed. Proper protocols should be used to sample and ship moldy grain for mycotoxin analysis.

Key management steps to minimize grain mold and mycotoxin contamination

- Ensure proper storage conditions — grain moisture, temperature, relative humidity
- Minimize mechanical damage — harvest and postharvest shipping and handling
- Minimize insect damage — pre-harvest and postharvest storage
- Plant tolerant hybrids — some tolerant hybrids available
- Sanitation of storage facility — critical management practice
- Chemical management — propionic acid, mineral oils
- Assay moldy grain for mycotoxins — Toxicology Lab, Vet Diagnostic Center, UNL Segregate, blend, or destroy contaminated grain — as per FDA regulations

Unresolved Issues for Grain Molds and Mycotoxins

Several issues related to molds and mycotoxins remain unresolved:

1. The impact of mycotoxin ingestion on human health,
2. The importance of cropping practices (e.g., plant density and spacing, tillage, crop sequence) to the incidence and severity of grain mold, and
3. The identification of true grain mold resistance in hybrids.

REFERENCES

Bacha, H., Hadidane, R., Creepy, E. E., Regnault, C., Ellouze, F., Dirheimer, G. 1988. Monitoring and identification of fungal toxins in food products, animal feed and cereals in Tunisia. J. Stored Prod. Res. 24:199 – 206.

Bamburg, J.R., Strong, F.M., Smalley, E.B. 1969. Toxins from mouldy cereals. J. Agric. Food Chem. 17:443–450.

Blaney, B.J., Moore, C.J., Tyler, A.L. 1984. Mycotoxins and fungal damage in maize harvested during 1982 in Far North Queensland. Aust. J. Agric. Res. 35:463–471.

Bottalico, A., Visconti,.A., Logrieco, A., Solfrizzo, M., Mirocha, C.J. 1985. Occurrence of zearalenols (diastereomeric mixture) in corn stalk rot and their production by associated *Fusarium* species. Appl. Environ. Microbiol. 49:547–551.

Campbell, J.C., Stoloff, L. 1974. Implications of mycotoxins for human health. J. Agric. Food Chem. 22:1006-1015.

Charmley, L.L.; and Prelusky, D.B. 1994. Decontamination of *Fusarium* mycotoxins. In: Mycotoxins in Grain: Compounds other than aflatoxin. J.D. Miller and H.L. Trenholm (Eds.). Eagan Press, St-Paul, Minn. pp. 421-435.

Chelkowski, J. 1991. Mycological quality of mixed feeds and ingredients. In: Chelkowski, J. (ed) Cereal grain, mycotoxins, fungi and quality in drying and storage. Amsterdam : Elsevier, 1991, 217-227

Chelkowski, J., Trojanowska, K., Wiewiorowksa, M. 1983. Mycotoxins in cereal grains, Part VIII: Microbial evaluation of cereal grain quality connected with mycotoxin occurrence. Nahrung 27: 311–318.

Christensen, C.M., Sauer, D.B.1982. Microflora. In: Storage of cereal grains and their products (Christensen CM ed), American Association of Cereal Chemists, St. Paul, pp 219- 240.

Deacon, J.W. 1980. Introduction to Modern Mycology, Blackwell Scientific Publications, London, pp. 100–102.

Diekman, M. A. & Green, M. L.1992. Mycotoxins and reproduction in domestic livestock. J. Anim. Sci., 70:1615-1627.

Di-Menna, M.E., Lauren, D.R., Smith, W.A.1991. Effect of incubation temperature on zearalenone production by strains of *Fusarium crookwellense. Mycopathologia*, 116: 81–86.

D'Mello, J.P.F. 2000.Antinutritional factors and mycotoxins. *In* J.P.F. D'Mello, ed. *Farm animal metabolism and nutrition*. Wallingford, UK, CAB International. p. 383-403.

D'Mello, J.P.F. & Macdonald, A.M.C. 1998. Fungal toxins as disease elicitors. *In* J. Rose, ed. *Environmental toxicology: current developments*. Amsterdam, the Netherlands, Gordon and Breach Science Publishers. p. 253-289.

Dorner, J.W., Cole, R.J., Diener, U.L.1984. The relationship of *Aspergillus flavus* and *A. parasiticus* with reference to production of aflatoxins and cyclopiazonic acid. Mycopathologia 87:13-15.

Frazier,W.C., Westhoff, D.C.1978. Food Microbioogy 3 rd ed, Tata McGraw – Hill Publishing Company Ltd., New Delhi, pp. 454-463.

Hesseltine, C.W., Vandegraft, E.E., Fennel, D.I., Smith, M.L., Shotwell, O.L . 1972. Aspergilli as ochratoxin producers. Mycologia 64: 539 – 544. Hudson , H. J. (1986) Fungal Biology, Edward Arnold (Publishers) Ltd. London, pp. 180-182

Jimenez, M., Mateo, R., Querol, A., Huerta, T., Hernandez, E.1991. Mycotoxins and mycotoxigenic moulds in nuts and sunflower seeds for human consumption. Mycopathologia, 115: 121-127.

Marasas, W.F.O.; van Rensburg, S.J.; and Mirocha, C.J. 1979. Incidence of *Fusarium* species and the mycotoxins, deoxynivalenol and zearalenone, in corn produced in oesophageal cancer areas in Transkei. J. Agric. Food Chem. 27:1108-1112.

Merino, M., Ramos, A, J, Hernandez ,E. 1983. A rapid HPLC assay for zearalenone in laboratory cultures of *Fusarium graminearum*. Mycopathologia 121: 27-32.

Miller, J.D. 1994. Epidemiology of *Fusarium* ear diseases of cereals. In: Mycotoxins in Grain: Compounds other than aflatoxin. J.D. Miller and H.L. Trenholm (Eds.). Eagan Press, St-Paul, Minn. pp. 19-35

Opeke, L.K. 1982. Tropical tree crops, Spectrum Books Ltd., Ibadan, pp. 124-174

Patterson, J.P., Roberts, B.A. 1979. Mycotoxins in animal feedstuffs: sensitive thin layer chromatographic detection of aflatoxin, ochratoxin A. sterigmatocystin, zearalenone and T-2 toxin. J. Assoc. Off. Anal. Chem. 62:1265-1267.

Placinta, C.M., D'Mello, J.P.F. & Macdonald, A.M.C. 1999. A review of worldwide contamination of cereal grains and animal feed with *Fusarium* mycotoxins. Animal Feed Science and Technology, 78:21-37.

Prelusky, D.B.; Rotter, B.A.; and Rotter, R.G. 1994. Toxicology of mycotoxins. In: Mycotoxins in Grain: Compounds other than aflatoxin. J.D. Miller and H.L. Trenholm (Eds.). Eagan Press, St-Paul, Minn. pp. 359-403

Purseglove, J.W. 1974. Tropical Crops: Dicotyledons, Longman, Singapore, pp. 564-570

Richardson, K.E., Hagler, W.M., Mirocha, C.J.1985. Production of zearalenone, a – and b – zearalenol by *Fusarium* spp. in rice culture J. Agric. Food Chem. 33:862-866.

Robb, J., Norval, M. 1983. Comparison of cytotoxicity and thin – layer chromatography methods for detection of mycotoxins. Appl. Environ. Microbiol. 46:948-950.

Schnurer, J. - Jonsson, A.1992. Ergosterol levels and mould colony forming units in Swedish grains of food and food grade. In: Acta Agric. Scand., Sec. B, Soil Plant Sci.. 42:240-245.

Seiler, D. A. L.1986. Baseline counts for wheat, flour and bran. In: Williams, D. T. 1991. Storage of grains and seeds. pp. 721-746. In: Smith, J.E. - Henderson, R. S. (eds) Mycotoxins and animal food. London : CRC Press, s.194-197.

Smith, J.E., Moss, M.O. 1985. Mycotoxins: Formation, Analysis and Significance, John Wiley & Sons, New York.

Tanaka, T.; Hasegawa, A.; Yamamoto, S.; Lee, U.S.; Sugiura, Y.; and Ueno, Y. 1988. Worldwide contamination of cereals by the *Fusarium* mycotoxins nivalenol, deoxynivalenol, and zearalenone. 1. Survey of 19 countries. J. Agric. Food Chem. 36:979-983.

Ursula, B, Mirocha, C.J., Wen, Y. 1992. Production of zearalenone, monliformin and trichothecenes in intact sugar beets under laboratory conditions. Mycopathologia 119:167-173.

Wicklow, D.T., Shotwell, O.L. 1983. Intrafungal distribution of aflatoxins among conidia and sclerotia of *Aspergillus flavus* and *A. parasiticus*. Can. J. Microbiol. 29: 1-5.

Williams, D. T.1991. Storage of grains and seeds. In: Smith, J. E. - Henderson, R. S.: Mycotoxins and animal food. London : CRC Press. 721- 746

5

IPR Issues — Scenario of Indian Universities and Institutions

K.A. Anu Appaiah

Abstract

Majority of basic research in our country has foundations in Universities and various Institutions linked to them. This paves way for discovering wilder varieties of the present known cereals and other food crops. Such findings help in bioprospecting the required gene/genetic material for the improvement of the existing crops. As per the Biodiversity Act of 2002, it is the responsibility of these institutions to document the same. But, as research scientists are geared to meet the challenges? Being used to socialistic system from the days of Independence where knowledge generated was open to everyone or closed and kept under the mountains of files, the TRIPS agreement has come as a shock. But the true impact of it was realized in 1999-2000, when the patent amendments were taken up. The private institution/ companies went into the act immediately. We in the University neither were slow to react nor were we prepared to take the shock. In this paper some of the problems pertaining to IPR and scenario of Indian Universities and Institutions in this regard were highlighted.

INTRODUCTION

Majority of basic research in our country has foundations in Universities and various Institutions linked to them. This paves way for discovering wilder varieties of the present known cereals and other food crops. Such findings help in bioprospecting the required gene/genetic material for the improvement of the existing crops. As per the Biodiversity Act of 2002, it is the responsibility of these institutions to document the same. But, as research scientists are we geared to meet the challenges? Being used to socialistic system from the days of Independence where knowledge generated was open to everyone or closed and kept under the mountains of files, the TRIPS agreement has come as a shock. But the true impact of it was realized in 1999-2000, when the patent amendments were taken up.

The private institution/ companies went into the act immediately. We in the University neither were slow to react nor were we prepared to take the shock. In this paper some of the problems pertaining to IPR and scenario of Indian Universities and institutions in this regard were highlighted.

- Lack of knowledge on IPR
- Insensitive to the present requirements
- Complexity of legal issues in our country
- What to patent and what not to patent
- Criteria for patenting
- Fear of the unknown, hence objection for the change
- Financial implication
- Incomplete/missing local database
- Myths and belief

Thus it is the need of hour that discussion at various levels and sensitisation to the issues of knowledge related to IPR at college and University levels should be given a high priority. This would go a long way to bring us at par with the western set-up.

With the dawn of the new millennium India entered into the barely understood terrain of global patents regime. It was at that time people questioned whether our government would get its act ready. The questions posed at that time, were:

1. Can policies for confirming to the new regime be created?
2. Can the process of granting patents will be speeded?
3. Can patents application fee be a source of income?
4. Will there be penalties for violation of patents law?

Today we find that the two amendments to IPR and patent law the government has got its acts together. But we at the University and its institutions linked to them have we got our acts together? Unfortunately the answer is No.

In this paper it is tried to discuss a few of the issues which has made these institutions of higher education of not being equipped to the task related to IPR. A larger percentage of teachers are not aware of IPR nor any concrete efforts being made even in Orientation / Refresher courses to inform the teachers about this issues. UGC has no data on its web about the patent, copyrights or IPR related issues. Hence, it is a general apathy that these IPR related issues are only for R &D Institutions and companies. With funding for higher education becoming scarce, the question of financial implications looms large.

Indian society from time immemorial has been regarded knowledge as the highest virtue of man. The knowledge power is the determinant factor for progress. Our University set up with its large resource is the right place to start the Documentation of our vast resources. This is the first requirement of Biological diversity Act 2002. Till December of 2003 UGC has recorded 207 Universities, 5 institutions and 13 institutes of national importance and 88 deemed Universities. From a total of 313 organizations of higher education, 50 are considered to be centres of excellence. Leaving these centres of excellence the rest of the organizations have really fallen back when it comes to IPR related issues. India is the home for 8% of the recorded species in the world, has two hot spots of Biodiversity in the world. Hence the task of documenting is the duty of the University. We do agree University have their own limitations like-

1. Traditional set up of rules and regulation and status leaving little scope for quick improvement.
2. Academic structure is rigid, teaching and evaluation method outdated.
3. Work culture is normally static and rigid functioning compared to bureaucratic methods of functioning.
4. Quality and innovation and research have low social geographical and educational relevance.
5. University most often are examination-conducting bodies.
6. Faculty and course do not match with qualitative mechanism for improvement and knowledge advancement.

Kothari educational commission in its report wrote " there is a general feeling in India that the situation in higher education is unsatisfactory and even alarming in some ways that the average standards have been falling with that rapid expansion of educational institution has resulted in lowering of quality". An insight into the break-up of students opting for research in these institutions is just 1.5%. Added to this, there is a feeling in academic circles that diversion of efforts of a teacher towards R & D usually sacrifices teaching interest. But few realize that R & D leads to financial independence, means of recognition in the profession. Today, R& D is a valid means of Technology Development with these resources.

Until 1999 there was a general apathy towards the patenting. Data of 1998 shows that only 5,700 patents application were filed in India against 67,000 in China, 46,000 in South Korea and 26,000 in Brazil. Of these, only 16 were from University and its allied institutions. Today the situation is different with more than 26,25 patent applications filed till September 2007. But the paper processing for patent in our country still takes 3-4 years compared to 18 months in U.S.A.

Universities in general pursue research which many times has no relevance for its surrounding society, hence gets alienated. This also makes Industries to lean to other R & D centres for their requirements.

In general our society is yet to wake upto the fact that "Knowledge is Money" hence payment for advice's at local /lower levels are still a misunderstood concept. Returns from patents are still insignificant (Table 1). This leads to very low off-take of patented materiel for production which is compounded by our complex legal aspects. If and when IPR issues are taken seriously R&D gets affected in the following ways.

1. Research results must be patented before publication.
2. Corporate & laboratories will collaborate on R&D
3. Commercialisation of research will earn royalties to fund R&D
4. Technologies developed will be licensed.

Table 1: Patent situation of India compared to other countries

Country	Patents / Million persons	Royalties earned / person
India	6.04	0.1
U.S.A	701.08	135.5
Japan	2875.68	82.4
China	71.60	0.1

Web data 2005, have to be updated

The concept of biodiversity and its related IPR should move beyond scientists and technologists to ordinary people. This is possible only with a proactive University in the local area. The equitable sharing of benefits arising out of the use of biological resource and knowledge relating them would make this proposition look bright. Benefits may include transfer of technology, location of research and development, association of local institutions and local people/ community. Such associations may lead to setting up venture capital funds, direct monetary compensation and other non-monetary benefits.

REFERENCES

Anonymous (2003). Battlelines drawn. Down to Earth Society for environmental communications 18.

Educational statistics (1999). GOI- MHRD.

Srinivas, M.R.S., Anu Appaiah, K.A. and Rao, K.S. (1994) Approach to scientific research – A personal guide. RSF, CFTRI, 71-75.

Upadhya (2003). Higher education under the WTO Regime. Univ. News 41 (31) 123-126.

www.aiuweb.org

www.did.co.in

www.patentindia.com

SECTION-II

QUALITY EVALUATION OF GRAINS AND ITS CONSTITUENTS

6

Adulteration in Cereal Grains: An Overview

G.S.Padda, Amarjeet Kaur and Poonam A. Sachdeva

Abstract

Adulteration refers to a process by which the quality or the nature of the given substance is reduced by the addition of foreign or an inferior substance, removal of vital component, packing and storage under harmful conditions, insect infestation, poisonous material, undesirable chemicals and unintentional contaminants. All these change in food whether intentional or unintentional are covered under food adulteration. In cereal grains adulteration results by incidental contamination during the period of growth, harvesting, processing, transport, and distribution and results in various health hazard, sand, marble chips, earth and filth are also adulterants in grains. Pest and pesticides residues are also hazardous.

INTRODUCTION

Grain food includes cereals like rice, wheat, maize or corn, jowar, ragi and bajra. These are mostly staple diet of average Indian and consumed for calories and protein. Earlier people used to grow own cereal, pulses, and other foods in their own garden. So, far them deciding the quality of food they consumed was not a difficult matter. They were getting everything pure and they lived a healthier and longer life. In today's industrialized age, the common man has no such facility because he is no longer associated with producing and processing wheat he consumes. Although he has an infinite number of choices to make from but he is left to mercy of manufacturers and salesmen. In most western countries, people have organized themselves into consumer association to disseminate information and to safe guard their interests.

Consumer goods are constantly tested at laboratories controlled by these associations and the test results are made available to the public. In India, unfortunately there is still no strong consumer association to make voice of consumer heard and nor there is any facility for comparative testing or informative labeling of consumer goods. Consumer is at the mercy of trader and manufacturer who is playing with the health, as number of food stuffs in the market are adulterated. Government records show that on an average 25-30% of edibles sold in various parts of country are adulterated. The consumers like to get maximum quantity for as low a price as possible. The sellers must meet the needs of the buyers to be able to exist. This is a vicious cycle. When the price of the food production is higher than the price, which the consumer is prepared to pay, seller is compelled to supply a food product of inferior quality, thus adulteration occurs.

To check such malpractice, PFA was passed in 1954 and came in force from June 01, 1955. It required that food be pure, wholesome and honestly labeled. Adulteration is defined as a process by which the quality or the return of the given substance is reduced by one way of following:

- The addition of a foreign or the inferior substance.
- The removal of a vital component.
- Packing and storage under harmful conditions.
- Insect infestation.
- Poisonous materials.
- Undesirable chemicals.
- Unintentional contaminants.

Food Laws and Control System in the country help the consumer in two ways.

- By providing food safety.
- By ensuring fair-trade practices.

According to PFA the article is said to be adulterated, if:

1. The article sold by a vender is not of native, substance, or quality demanded by the purchaser.
2. If the article contains any other substance, which affects or if the article is so processed as to affect injuriously the nature, substance or quality thereof.
3. If any inferior or cheaper substance has been substituted wholly or in part for the article, so as to affect injuriously the nature or substance or quality thereof.

4. If any constituent of the article has been wholly or in part abstracted so as to affect injuriously the nature, substance, or quality thereof.
5. If the article has been prepared, packed or kept under sanitary conditions whereby it has become contaminated or injurious to health.
6. If the article contains any poisonous or other ingredient which renders its contents injurious to health.
7. If the article is obtained from a diseased animal.
8. If the article consists wholly or in part of any filthy, putrid, disgusting, rotten, decomposed or diseased animal or vegetable substance or is insect infected or otherwise unfit for consumption.
9. If the container of article is composed of any poisonous or deleterious substance which renders its content injurious to health.
10. If any colouring matter other than that prescribed and is in amounts nor whether the prescribed limits of variability is present in the article.
11. If the article contains any prohibited preservative or permitted preservative in excess of the prescribed limits.
12. If the quality or purity of the article falls below the prescribed standards or its constituents are present in quantities, which are in excess of the prescribed limits of variability.

Adulteration of food may endanger health if the physiological function of the consumer is affected due to either addition of a deleterious substance or removal of a vital component. Adulteration of cereal grains may be intentional or unintentional. Intentional is willing to increase the margin of profit. Incidental is usually due to ignorance, negligible or lack of proper facilities. Intentional adulterants are sand, marble, chips, stones, mud, chalk powder, water, mineral oil, and coal tar dyes.

Usually in wheat grain, adulteration is considered when the impurities, damaged, shriveled grain and foreign matter exceed the standards e.g. for Durum and soft wheat, maximum limits are prescribed in table 1.:

Table 1: Limits of impurities in wheat verities

Impurities (% max)	Durum wheat	Soft wheat
Damaged	6.0	6.0
Heat damaged, max	0.5	0.5
Broken & shriveled	0.8	0.8
Other cereals	3.0	3.0
Foreign matter	1.0	1.0
Harmful/ Toxic seeds	0.5	0.5

Adding of moisture is also a type of adulteration in wheat grain. Admixture of low-grade varieties in durum wheat is also a type of adulteration. Even mixing of bajra and ergot, dhatura seeds are quite common. Purple black longer size grains in bajra show the presence of ergots, even ergot floats at surface of glass containing 20% salt solution. Dhatura seeds resemble chilies seeds with blackish brown color, which can float over the surface grains while sound grains settle down.

In paddy main adulterant is moisture, which exceeds the permissible limits of 14 percent. Admixture of low-grade varieties in the basmati, exceeding the levels of brokens, foreign matter, and discolored, chalky, insect damaged grain are all adulterants. Specification of rice for export is given in table 2.

Table 2: Maximum acceptance limits for Rice verities

Specifications	Basmati	Non-Basmati
Broken content (%)	25	10
Foreign matter (%)	0.005	1.0
Damaged/ Discoloured (%)	0.025	2.0
Paddy grain (%)	0.1	0.1
Moisture (%)	14.0	14.0
Chalky /immature %	1.0	5.0
Admixture %	10.0	15.0
Insect damage %	Nil	Nil

Agmark has also laid down specification for export of Basmati and Non Basmati Rice as shown in table 3.

Table 3: Maximum limits for export of rice verities as per agmark

Grade	Maximum Limits (%)				
	Foreign Matter	Broken fragments	Other rice & red grain	Damaged discoloured	Moisture content & chalky grain
Special	0.5	5.0	10.0	1.0	14.0
A	1.0	10.0	15.0	2.0	14.0
B	2.0	10.0	20.0	3.0	14.0

Any deviation from these limits may be termed as adulteration.

Long, slender, highly fragrant Basmati rice is highly priced both locally and abroad. There are several varieties of ordinary, non-scented rice, which resemble Basmati in one or several physical characteristics. Adulteration of the highly priced, flavourful Basmati with cheaper varieties such as Pusa-109, Sharbamati, PR-106 and Parmal yields substantial profit to the traders. Basmati may be adulterated to the levels around 30-40%. So far no reliable qualitative and quantitative tests are available to detect adulteration of Basmati rice except a report on use of a kernel elongation test for this purpose. It is proposed that length breadth ratio of 3.92± 0.09 to 4.09 ± 0.09 could be taken as the reference value for pure Basmati and therefore length/ breadth ratio is a promising indicator of adulteration.

A variety of maize based food and snack items have been launched in the domestic as well as International Market. Grain should consist of 50% or more of whole kernels of shelled dent corn or shelled flint corn and not more than 10 percent other grains. Each determination of class, damaged kernels, heat damaged kernels, waxy corn, flint corn and dent corn is made on basis of the grain after the removal of the broken corn and foreign material. Similarly barley should not have any admixture of pesticide, fungicide and any obnoxious and toxic material. Other grains should not be more than 6%. Grain should not be having more than 8% of shriveled or broken kernels. In Ragi, not more than 4% foreign matter, 3% other food grains and 3% damaged grains and not more than 8% of immature grains are allowed.

Talc or soapstone and chalk powder are used to adulterate arrowroot powder, wheat flour atta, maida, etc. They are also used as polishing powder for pulses and rice. Talc, hydrated magnesium silicate, sand, stone, mud, earth etc. are also silicate, which the human system cannot digest. Moreover they may empower the normal smooth functioning of system in various ways. Varieties of talc or sand stone containing submicroscopic fibers of asbestos are more toxic than the rest, since these are believed to give rise to stomach cancer.

Adulteration from Heavy Metals

Grains become tainted with toxic metals such as lead, mercury, cadmium, tin, arsenic copper, zinc, antinomy chromium, cobalt through pesticide. These are permitted to some extent. But when the concentration exceeds the tolerate limit generally they produce a quick onset of symptoms such as vomiting, nausea, and pain in the stomach. Smaller amount may not cause any vomiting but only gastro intestinal disturbance. Arsenic pesticides (weed killers, fungicides, rodenticides, insecticides) are the main source of arsenic contamination in grains. Zinc comes from irrigated water. Strict control of various metal contaminants is desired: Limits for Heavy metal contaminants in food are given in Table 4 and that of pesticides/ insecticides residue in food grain in Table 5.

Table 4: Limits for heavy metal contaminants in food grains under PFA rules, 1955

S. No.	Characteristics	Limits (ppm)
1	Lead	2.5
2	Copper	30.0
3	Arsenic	1.1
4	Zinc	50.0
5	Cadmium	1.5
6	Mercury	1.0

Table 5: Limits of pesticides/insecticides residues (ppm) in food grains/milled food grains under PFA rules, 1955

	Characteristics	Food grains	Milled Food grains
1.	Aldrin, dieldrin, singly or in combination expressed as dieldrin	0.01	- -
2.	Carbaryl	1.50	- -
3.	Chlordane GS + TPANS	0.05	- -
4.	Diazinon	0.05	- -
5.	Dichlororos	1.00	- -
6.	Fenitrthion	0.02	- -
7.	Heptachlor	0.01	- -
8.	Hydrogen Cyanide	37.5	3.0
9.	Hydrogen phosphide	0.05	0.01
10.	Inorganic bromide	25.0	25.0
11.	Lindane	0.25	- -
12.	Malathion	4.0	1.0
13.	Phosphamidon	0.05	- -
14.	Pyrethins	1.50	0.5

It can be concluded that "Beware of these" common adulterant and as being an educationist it becomes our responsibility to make general public aware of these adulterant and their toxic effects on body.

7

Quality Evaluation of Bakery Flour

Amarjeet Kaur, G.S. Padda and Umesh Kamble

Abstract

Determination of quality of wheat is very important because quality requirement of an individual product vary from one another. Flour quality can be defined as the ability of the flour to produce a uniformly good end product. Flour quality factors can be divided into two basic groups: those that are inherent in the wheat as a result of the genetic make up and the changes brought about by growing conditions; secondly those factors that might be altered during the process of converting wheat into flour Wheat flour constituents play an important role in determining the quality of flour. The flour quality for baking could be assessed by measuring the physical, chemical and rheological characteristics.

INTRODUCTION

The annual production of wheat based products particularly breads, biscuits and other bakery items in India is estimated as 30 million tones. There are more than 75 thousand bakery units in urban and rural India. With the increase in population, popularity and demand & safety of bakeries and confectionary products, a great impetus has been witnessed in the development of new additives, ingredients and latest manufacturing technologies. In general bakery industry can however, be categorized as bread, biscuits and cake industry (Table 1).

Table 1. Classification of Baked Products

Sl. No.	Type of Leavening	Examples	Functions
1.	Unleavened	Pie crust, structured, tortilla, chapaties, parontha, etc.	Do not use leavening agent and shrink during sheeting.
2.	Leavened by water vapour	Puff pastry, crisp bread, Cream puff, popovers etc.	Leavening is achieved by water vapour.
3.	Leavened by air only	Angle food cakes, poured Cakes, sponge cakes, Cheese cakes, etc.	Air cells are produced during mixing of materials.
4.	Chemically leavened	Soda bread, muffins, cookies pancakes, snack crackers, layer cakes, & doughnut.	Air cells are produced by adding chemical leavening agents during mixing.
5.	Yeast leavened	Whole wheat bread, rye bread, sour bread, Danish pastry, doughnuts, cookies.	Addition of yeast in dough produces carbon dioxide, which acts as leavening agent.

All sorts of wheat flour are not suitable for bakery products. Quality of wheat flour for bakery and confectionary products is given in Table 2. There are basically three types of wheat varieties: *Triticum vulgare, Triticum durum* and *Triticum compactum*. Bulk of wheat is of the "*Vulgare*" variety used for manufacture of bread and biscuits. The "durum" variety of wheat is used to manufacture pasta whereas very soft for biscuits (*compactum*).

Table 2: Quality of Wheat Flour

Sl. No.	Quality of wheat flours	Uses
1.	Wheat flour high in protein (11-14%) content and strong in gluten.	Good for making yeast leavened bread and rods.
2.	Wheat flour intermediate in protein (8-11%) content and gluten strength.	Satisfactory in making yeast leavened white bread, cakes.
3.	Wheat flour low in protein (8-8.5%) content and weak gluten.	Used for making cakes and pastries.
4.	Lowest protein (7-8.5%) content and weak gluten.	Used for cakes and cookies.
5.	High protein (12-14%) content durum	Pasta products spaghetti noodles, macaroni etc.

Wheat mainly consists of three major components i.e. germ, endosperm and bran. These constituents are 2%, 85% and 13%, respectively. The chemical composition of these constituents will demonstrate the chemical role of these constituents (Table 3) during manufacturing as well helps the flour mills to manufacture the quality of flour required by bakery product manufacturer.

Table 3: Chemical Composition of Three Major Constituents of Wheat

Chemical Composition (%)	Major Constituents		
	Bran	Germ	Endosperm
Moisture	8-9	9.2	14.0
Protein	15-18	28.9	9.6
Fat	4-5	9.7	1.4
Carbohydrate	61-73	46.0	74.3
Fibre	2-13	2.1	—
Ash	3-6	4.1	0.7

The wheat composition consists of average composition of all these constituents. Approximate composition of wheat is given in Table 4.

Table 4: Approximate composition of wheat

Moisture, %	9-18
Starch, %	60-68
Protein, %	8-15
Cellulose, %	2-2.5
Fat, %	1.5-2.0
Sugar, %	2-3
Ash, %	.5-2.0

The wheat flour that is used for manufacture of various bakery products mainly consists of the endosperm. To understand bakery operation in relation to the constitution of wheat flour, there are two major elements viz. starch and gluten (protein), which affects the quality, and yield in bakery operation. It is important to understand the chemical composition and their all-level effects in the processing.

The profitability of flourmills is dependant upon the extraction rate. High extraction provides better contribution and thereby increases overall profitability. Higher extraction produces wheat flour of poor quality and also varies in composition. It affects the gluten quality. Higher extraction rate damage the starch quality. The variation in composition due to extraction rate is given in Table 5.

Table 5: Composition of Wheat Flour

Composition (%)	72% Extraction	80% Extraction
Moisture	13-15.5	13-15
Starch	65-70	64-69
Protein	8-13	9-14
Cellulose	0-0.2	0.2-0.3.5
Fat	1.5-2.0	1.5-2.0
Ash	0.3-0.6	0.6-0.8

Extraction rate about 72% cannot be used for bakery because of contamination of bran particle, which darken the flour color. Bureau of Indian Standard has laid specification of wheat flour for use in the bakery Industry (Table 6, 7). Wheat flour (maida) shall be the product obtained by milling cleaned, hard or soft wheat or blends there of in roller flour mill and bolting. The flour should be free flowing, dry to touch, should not pack when squeezed, should be creamy in colour and free from any visible bran particles. The flour shall have a characteristic taste and smell and shall be free from insect, rodent or fungus infestation, dirt and other extraneous matter. It shall also be free from fermented, musty or other objectionable odour. It shall not have any musty flavour or rancid tars.

When deemed necessary by the matter, the flour may be treated with bleaching agents and improvers allowed under PFA Rules not exceeding the limits specified against each, which are as follows:

Benzoyl peroxide – 10 ppm/(Max.)
Potassium Bromate – 20 ppm/(Max)
Ascorbic acid – 200 ppm/(Max)
Chlorine – 2000 ppm/(Max)

The product should conform to the requirements given in Table 6 & 7.

Table 6: Microbiological Requirements of Wheat Flour for use in the Bakery Industry

S. No.	Characteristics	Requirement
1.	Total Bacterial count per g Max.	100000
2.	Coliform bacterial count per g Max.	500
3.	*E. coli*	Absent in 1g
4.	Yeast / mould count per g Max.	10000
5.	Bacterial rope spore /g Max.	100

Table 7: Requirements of Wheat Flour (maida) for Use in Bakery Industry

S. No.	Characteristics	For biscuit	For bread	For cake
1.	Moisture, % by mass, Max.	13.0	13.0	13.0
2.	Total ash (d.b.), % by mass, Max.	0.5	0.5	0.5
3.	Acid insoluble ash (d.b.), % by mass, Max.	0.05	0.05	0.05
4.	Gluten (d.b.), % by mass, Min.	7.5	9.5	7.9
5.	Total protein (d.b.), % by mass, Min.	9.0	11.0	9.0
6.	Alcoholic Acidity (d.b.), % by mass, Max.	0.10	0.10	0.10
7.	Water absorption, %, Min.	55	60	-
8.	Maltose, %	-	2 to 3	-
9.	Sedimentation value, Min.	22	22	-
10.	Granularity	To satisfy test	To satisfy test	To satisfy test
11.	Aflatoxin Mg/ Kg Max.	30	30	30
12.	Uric acid mg/100g Max.	10	10	10

Assessment of Quality of Flour

I. Physical characteristics

Colour: The colour of flour depends on: a) Extraction rate of flour, b) Amount of pigment and, c) Flour particle size. The dullness or brightness of the flour is due to contamination by bran particles. Higher the flour extraction rate duller the colour and vice versa. The colour of the flour could be measured objectively by Kent Jones and Amos colour grader. Wheat of different origin, growing conditions and variety yield flours of different degree of yellow pigments. The millers of bread flour have problems in maintaining minimum yellow pigmentation. The coarse flour generally looks dull and darker than its finer counterpart owing shadow effects of larger particles.

Particle size distribution: Particle size distribution tests confirmed that flour is a homogenous material. The flour hydration rate determines the amount of water needed to the optimum dough finer particles accelerates faster hydration.

II. Chemical characteristics

Moisture: The moisture content of the flour is influenced by tempering condition of wheat prior to milling, duration of storage, temperature, and relative humidity. Higher Moisture leads to deteriorative chemical changes. The PFA limits for moisture are 14% and for BIS it is 13%.

Ash: The ash content in flour depends on the extraction rate of the flour. Higher the extraction rate, higher is the ash content because of the inclusion of branny layers rich in mineral and these tend to decrease keeping quality of the flour, which is also effected due to the presence of higher level of lipolytic enzymes in the high extraction rate flours.

Acid insoluble ash: The water insoluble impurity matter like silicates, sand etc. is indicated acid insoluble ash, which should not exceed 0.05%. This affects the quality and is considered as health hazardous.

Protein content and quality: Wheat is a unique grain when compared to other cereals. Wheat flour forms cohesive and elastic mass when treated with right proportion of water. This property is attributed to presence of gluten, which has the gas holding property during fermentation and setting to a spongy texture on baking. This is the most single constituent of the wheat flour, which affects the quality of the product. Quality wise the protein may range from 6-19%. However, gluten-forming proteins are present only in endosperm portion of wheat kernel.

Protein quality is determined by the classic Kjeldahl nitrogen analysis. Another method to determine wet and dry gluten content of flour. Flour used for bread should have higher gluten content ranging from 10 to 15%. While biscuit and cake flour require lower gluten content ranging from 7 to 10%. Generally, gluten absorbs two times its weight of water. Protein quality with reference to baking mostly refers to the quality of the gluten proteins, which consists of glutenin and gliadin contributing to extensibility and elasticity of the dough. Good quality protein with high extensibility and elasticity is needed for bread. For biscuits protein quality refers to low elasticity and high extensibility. The protein quality is evaluated by the following methods.

a. ***Zeleny sedimentation test:*** In this test the flour is treated with dilute lactic acid and allowed to sediment. Flour with strong protein gives higher value. Accordingly, sedimentation value ranges less than 20 for soft wheat flour, 20-40 for medium hard and more than 40 for hard wheat flour. The modified test using sodium dodecyl sulphate (SDS) gives consistent and reliable values even in the presence of high alpha amylase activity.

b. ***PelshenkeTest:*** This is measured by the time taken for a dough ball made from flour, yeast suspension to disintegrate at constant temperature of 30°C soft a biscuit flour disintegrate much faster as compared to bread flour.

c. ***Dough raising capacity:*** The expansion of dough made by mixing flour, yeast, and sugar in a cylinder for a period is indicative of its gas retaining property. The extent of expansion depends on both quality and quantity of proteins present in the flours.

***Carbohydrates**:* The principle carbohydrates are starch, pentosans, cellulose, and sugars. Generally, the cellulose fraction i.e. bran is removed during milling.

a. ***Starch**:* Starch forms the main constituent of the flour accounting for75-80% of the total weight. The starch components are mainly amylose (20-25%) and amylopectin (77- 80%). Starch is important in dough formation from its view of structure building as it influences the dough rheology and water absorption capacity phenomenon in bread.

b. ***Damaged starch**:* some fraction of starch gets damaged during milling i.e. starch granules loose some extent their intactness. The damaged starch differs widely when compared to native intact starch. 1g of the native starch absorb 0.4g of water while 1 g of damaged starch absorbs 2g of water. These damaged starch are more prone to amylolytic enzyme action, which produce different sugars. This damaged starch has a significant role to play in the fermented dough products and desired level is 7-9%.

c. ***Sugars**:* Level of sugar ranges from 1-2%. These are important in initial fermentation.

d. ***Pentosans**:* These are made up of xylose and arabinose and are only 2-3%. These influence the water absorption capacity. 1g of pentosan can absorb 1.5g of water and addition of pentosan can impart rigidity to the dough.

Lipids: Lipid content varies from 1-2%. These are involved in the formation of gluten network and thus help in gas retention. Free polar lipids improve the loaf volume even if they are in small amounts.

Enzymes: Most important types of enzymes present in wheat flour are proteolytic, lipolytic, and amylolytic. These enzymes are more important in fermented type of products by their mellowing action. Lipoxidase strengthens the dough and helps bleaching the flour.

Determination of alpha amylase activity

a) ***Amylograph method:*** Using Brabender amylograph the changes in viscosity of flour and water suspension are measured during a standard heating cycle, which gives the gelatinizing temperature, and peak viscosity. The value for normal flour is ~ 450 BU.

b) ***Falling number test:*** This gives idea about the alpha amylase content. Higher value is indication of low amylase. And value of around 300 is desired.

Determination of diastatic activity: The determination gives information regarding combined activity of diastatic enzyme. A value of about 2.5 to 3% of reducing sugar is normal for bread flour in 1 hour at 30°C.

Water Absorption: It is measured as the amount of water required to yield a batter or dough of desired consistency. Water absorption is influenced by gluten, damaged starch, pentosans and minor constituents such as dextrins, cellulose and flour particle size.

III. Rheological Characteristics

Flow properties of dough are measured by instrument and data on dough properties is used to product suitability of flour for various bakery products. The instruments are categorized on the basis of their function.

1. Recording dough mixer: Brabender farinograph, mixograph
2. Load extension meter: Extensograph, Chopin alveograph
3. Extension meter: Research water absorption meter
4. Viscometer: Hagberg's falling number, Brabender viscoamylograph
5. Gas production or gas retention instruments: Expansograph, Brabender maturograph, Brabender oven rise recorder.

It can be concluded that determination of quality of wheat is very important because quality requirements of an individual product vary from one other. Physical test, chemical test and electrophoretic techniques are used to segregate good quality. Rheological determinate helps to study dough properties and products the behaviour of flour during the baking process.

8

Functionality of Wheat Gluten Proteins

Bhupendar Singh Khatkar

Abstract

Wheat gluten consists mainly of gliadin and glutenin polypeptides. During mixing gliadin and glutenin proteins interact to form a unique viscoelastic gluten network, which is envisaged as being necessary for holding the gases and for producing a light porous crumb textured bread. Recent work has confirmed that the elastic properties of gluten are due to the glutenin fraction, whilst the viscous properties are attributed to the gliadin fraction. Equal amount of these two major protein components of wheat gluten is required for achieving the desired bread quality, but variation in the composition and physical properties of the glutenin polypeptides appear to be largely responsible for the differences in the gluten viscoelasticity and breadmaking potential among wheat cultivars. Recently, associations of gliadin polypeptides with breadmaking quality have also been reported. Using improved protein separation and purification techniques, physical methods and genetic engineering, we are beginning to understand the structure-functional relationship of wheat gluten proteins, but much remains to be explored in the years to come.

INTRODUCTION

Wheat is cultivated in all parts of the world except Antarctica. A world wheat area is around 220 million hectares and its production is around 550-600 million tonnes per annum. Most wheat is consumed in the form of various types of baked products in most countries of the world. Good bread quality flour encompasses an optimum blend of all proximate components. However, the technological importance of wheat flour is attached mainly to its gluten proteins, i.e. the water-insoluble complex proteins. The paramount importance of gluten proteins among wheat constituents for bread making quality is attributed to its unique physical properties such as viscoelasticity, water absorption capacity, film forming ability and thermosetting properties. Recently, wheat protein research has gained considerable momentum to understand how wheat gluten acquires these special physical properties and also efforts have been made to understand how these properties help to retain gas in bread making process. Here relationship of physical properties and structural aspects of wheat gluten proteins with bread making quality will be discussed.

CLASSIFICATION OF WHEAT GLUTEN PROTEINS

Proteins of wheat flour can be separated into non-gluten forming and gluten forming groups, when wheat flour is wetted and mixed with water. The term gluten refers to a viscoelastic protein mass, which is recovered after aqueous washing out of starch and water soluble components from wheat flour, when dried it yields a cream coloured free flowing powder of high protein content (75-80%) and bland taste. When rehydrated, it regains its viscoelastic properties, i.e. rubber like characteristics. The non-gluten forming proteins are albumins and globulins, which account approximately 10-20% of the total flour proteins and their amounts are relatively constant among varieties.

According to Osborne, gluten-forming proteins, which represent 80-90% of the total proteins of wheat flour, have been classified into two major groups, viz., gliadin and glutenin, based on their extractability and unextractability, respectively, in aqueous alcohol. The terms, gliadin and glutenin, are the specific names for the wheat proteins, corresponding to the generic terms prolamin and glutelin of all cereal proteins. Neither of these groups are considered to consist of pure proteins and a considerable amount of overlap between these groups may occur, depending on the exact extraction conditions, such as time, temperature and type, as well as concentration of alcohol used. In view of this disadvantage of Osborne classification, an alternative classification of gluten proteins has been proposed that reflects biological, chemical and genetic relationships among component polypeptides of the gluten complex. They divided gluten proteins into three main groups, viz., sulphur-poor prolamins, sulphur-rich prolamins and high molecular weight (HMW) prolamins. All these gluten protein groups are considered typical prolamins, because they are rich in proline and glutamine amino acid residues and are extractable, at least partially in aqueous alcohol, particularly in the presence of reducing agent.

The gliadin and glutenin fraction can be divided into subclasses. The gliadins are usually separated by lactate polyacrylamide gel electrophoresis (PAGE) into four main sub-categories, alpha-, beta-, gamma- and omega-gliadins in decreasing order of electrophoretic mobility under acidic conditions and increasing order of relative molecular mass. Amino acid sequencing has revealed that the alpha- and beta-gliadins are structurally closely related polypeptides and, therefore, both of these gliadin polypeptides have been classified into one group, i.e. alpha-type gliadins. The relative molecular masses of alpha-type and gamma-gliadins are between 30 000-45 000, whereas omega-gliadins generally have relative molecular mass of 44 000-80 000. The gamma-gliadins, in general, have somewhat higher relative molecular masses than alpha-type gliadins.

According to their sodium dodecyl polyacrylamide gel electrophoresis (SDS-PAGE) under reduced conditions, the glutenin fraction can be divided into two sub-categories, i.e. low and high molecular weight subunits. The high molecular weight (HMW) subunits, also known as A subunits of glutenin are the largest polypeptides and contain the highest level of glycine residues among gluten proteins. Their molecular weights measured by SDS-PAGE range from 80 000-160 000. The HMW subunits are further divided into x-type and y-type. The x-type subunits are somewhat large polypeptides, while the y-type subunits are smaller.

STRUCTURAL PROPERTIES

In recent years, our knowledge concerning the structure of gluten proteins has improved enormously. The amino acid sequences of purified individual gluten proteins are now available. Comparison of these sequences reveals that all types of gluten protein polypeptieds, in general, have at least two or three distinct structural domains, i.e., a central repetitive domain, flanked by non-repetitive C-terminal and N-terminal domains. The S-rich prolamins are characterized by a shorter repetitive domain and a longer C-terminal domain compared with HMW prolamins and S-poor prolamins. S-rich prolamins have their cysteine residues in the C-terminal domain, the only exception being the LMW subunits of glutenin, which also contain one cysteine residue in the N-terminal domain, S-poor prolamins have no cysteine residues, as indicated from omega-secalin of rye, a homologue of wheat omega-gliadins. On the contrary, HMW prolamins have the majority of their cysteine residues (3 and 5 in x-type and y-type subunits, respectively, in the N-terminal domains and only one in the C-terminal domain. Several y-type subunits and x-type subunit 5 contain an additional cysteine residue in the central repetitive domain.

Associations of Wheat Gluten Proteins with Mixing Properties and Bread Making Quality

It has long been recognized that the quantity and quality of their proteins govern the mixing properties and bread making quality of wheat flours. Recent findings indicate that the mixing characteristics of flours from wheat cultivars in the medium-strong range are significantly influenced by their gluten protein contents. On the other hand, the mixing properties of 'extra strong' or weak flours are relatively less affected by their gluten protein contents, but gluten protein quality primarily controls their behaviour during mixing. The weak gluten develops quickly, where as strong and 'extra strong' glutens need longer time to mix to peak dough development (PDR). This clearly suggests that the mixing requirement of flour is related to its gluten protein composition. It has been demonstrated that gliadin fraction generally decreases mixing time, whereas certain glutenin fraction increases the mixing requirements of wheat flour. Addition of different gliadin subgroups to a base flour decreases mixing time in the order omega->gamma->beta->alpha-gliadins in accordance with the decrease in their molecular size and \ or the increase in their charge.

The type and composition of HMW glutenin subunits have also been considered important for determining the mixing characteristics and bread making performance. It is generally considered that subunits 5+10, 17+8, 1 and 2* are related to longer mixing times, dough strength and superior bread quality, whereas subunits 6+8, 2+12, 3+12 and 20 are related to shorter mixing times and dough weakness. Both quantitative and qualitative effects may contribute to quality difference associated with specific HMW glutenin subunit alleles. The alleles at the *Glu-B1* and *Glu-D1* loci produce HMW subunits (17+18 and 20 or 5+10 and 2+12) in similar amounts. This indicates that the qualitative differences in HMW subunits may be responsible for the differences in the mixing characteristics and bread making quality among wheat cultivars. This is consistent with the previous observations that the superior subunit 5 has one additional cysteine residue and has a greater molecular mass than the inferior subunit 2. The superiority of the subunits 17+18 over subunit 20 in the case of genome B has been attributed to the number and position of the cysteine residues. Furthermore, on the basis of amino acid sequence differences in the regions towards the C-terminal ends of the repetitive domains of HMW glutenin subunits 10 and 12, it has been postulated that subunit 10 may form a more regular beta-spiral structure than subunit 12, and that this in turn may lead to improved elastic properties. Beta-spiral structure has been proposed to be intrinsically elastic.

Gluten Viscoelasticity and Its Association with Baking Quality

The dynamic rheological analysis differentiates glutens extracted from the cultivars into 'extra-strong', strong and weak based on their elastic moduli measured in the linear domain (Fig. 1). This classification is in accordance with the classification obtained using the 2g Mixograph. Qualitatively, the glutens show very similar rheological behaviour, i.e. elastic moduli increases moderately with increasing frequency, but, exhibits large quantitative differences in the magnitude of their elastic moduli (Fig. 1). Differences in the viscoelastic behaviour of the glutens are linked to the role of cross-links, the rheology of gluten sub-fractions, the effect of varying the gliadin/glutenin ratio, and the effects of the relative quantity and quality of glutenin high M_r subunits.

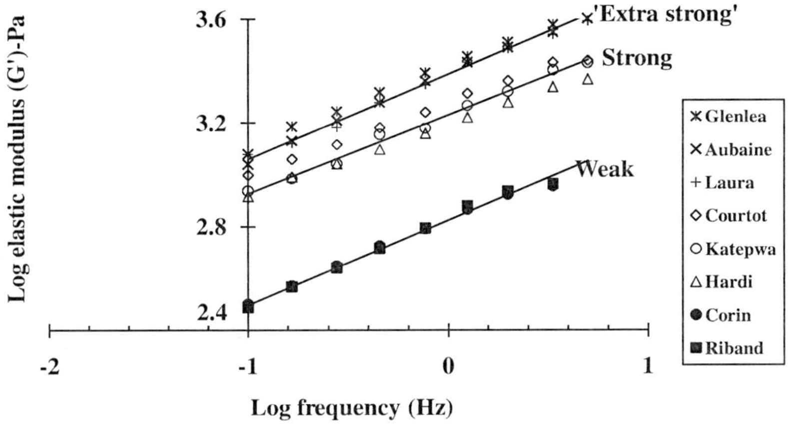

Fig. 1: Mechanical spectra of glutens from different wheat cultivars

'Cross-links', such as disulphide bonds, hydrogen bonds and hydrophobic interactions, have long been considered important in determining the rheological properties of wheat gluten. However, the relative importance of these cross-links to gluten viscoelasticity is still unclear, and their underlying mechanism of action in influencing gluten viscoelasticity also remains to be defined. The treatment of gluten with urea decreases the elastic characters of the 'extra strong' (cv. Glenlea) and weak (cv. Riband) glutens up to 3M concentrations, and thereafter it enhance the elasticity of these glutens (Fig. 2).

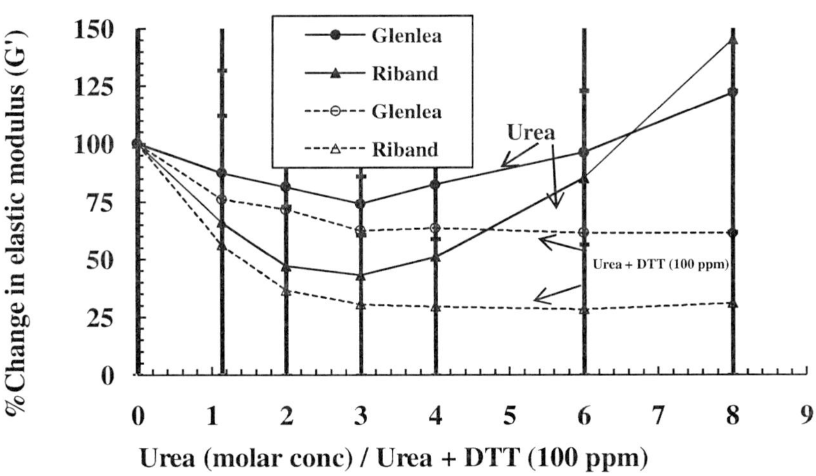

Fig. 2: Effect of urea and urea plus dithiothreitol on the elastic modulus of the 'extra strong' (cv. Glenlea) and weak (cv. Riband) glutens

The decrease in the elasticity of the glutens may be due primarily to cleavage of hydrogen bonds and, probably, hydrophobic interactions by urea. It is speculated that the enhanced elasticity of the glutens treated with higher concentration of urea solution (>3M) might be linked to the fact that the progressive disruption of non covalent interactions, i.e. hydrogen bonds and hydrophobic interactions, brings about unfolding of the gluten polypeptide chains, thus giving them freedom to adopt more extended conformations. This increment in the length of the gluten polymers is envisaged as providing conditions conducive to entanglement coupling.

Under these circumstances, elasticity is increased because longer times are required for the entangled polymers to move past each other when such a network is subjected to shear stress. It is also possible that unfolding of gluten polypeptide chains upon disruption of non covalent interactions exposes buried SH groups within the protein macromolecules, which may then react with S-S bonds by SH/SS interchange reactions to form new disulphide linkages, thus leading to an increase in the resistance to deformation. This appears conceivable on account of the fact that treating these glutens with urea in the presence of dithiothreitol (DTT) did not increase the elasticity of the glutens above 3M urea as was the case with urea alone.

It is noteworthy that the difference in the elastic character of the two glutens was at its greatest at 3M urea alone and at 3M urea in the presence of DTT. The elastic character of 'extra strong' gluten (cv. Glenlea) was significantly greater than that of weak gluten (cv. Riband), suggesting the greater importance of disulphide bonds in the structure of the 'extra strong' gluten from cv. Glenlea. This was further demonstrated by treating these glutens with increasing levels of DTT. Untreated glutens had very wide differences in their elasticity, i.e. the elastic character of 'extra strong' gluten (cv. Glenlea) was significantly greater than that of the weak gluten (cv. Riband), but the differences decreased progressively as the DTT concentration was increased (results not shown). It appears, therefore, that disulphide cross-links are important determinants of differences in elasticity between these glutens.

The rheological profiles of glutens and gluten sub-fractions from good (cv. Hereward) and poor (cv. Riband) bread making wheats showed significant differences in the magnitude of the dynamic moduli and their frequency dependence (Fig. 3). However, as far as it is known, the dynamic rheological properties of gluten sub fractions, i.e. gliadin and glutenin have not been investigated previously. In this study, the glutenin sub-fractions had slopes close to zero for G¢ (Fig. 3), which is typical of the behaviour of a viscoelastic solid. These spectra reflect the existence of a three dimensional network with junction zones having a longer lifetime than the observation time. On the other hand, the mechanical spectra of gliadins were typical of those of an elasto-viscous liquid, in which the viscous moduli are higher than the elastic moduli and in which both show a strong frequency dependence (Fig. 3). That is, they increase with decreasing observation times.

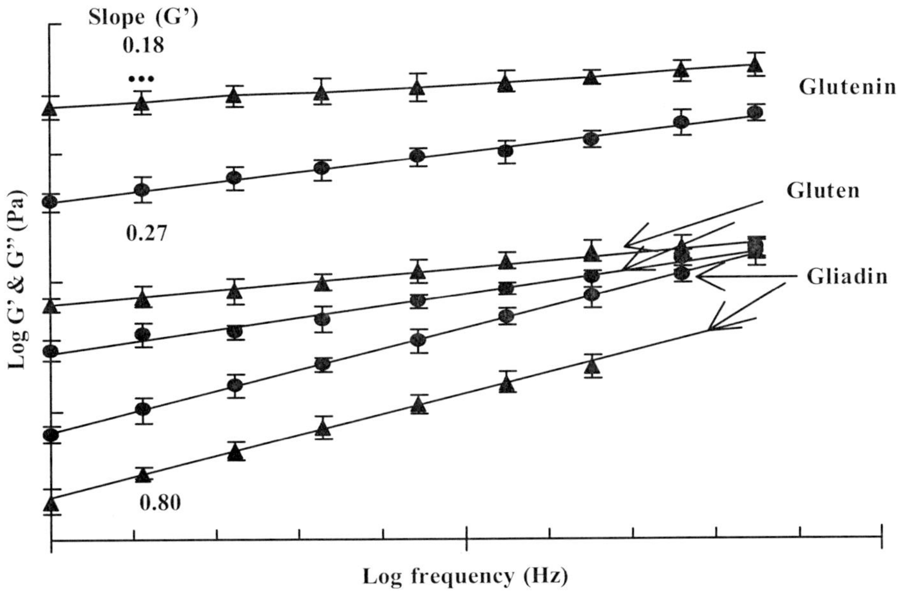

Fig. 3: Comparison of mechanical spectra of gluten and gluten sub-fractions from cv. Hereward. G¢ = triangles; G¢¢ = circles

This means that molecular motions occur within the experimental time scale and that there are no long time interactions between gliadin polypeptides, suggesting that gliadin has a high concentration of uncross-linked polypeptides. In summary, comparing the rheological profiles of glutenin, gliadin and gluten, it is evident that glutenin forms a highly cross-linked system, whereas gliadin forms a weak but highly viscous system with little cross-linking, with gluten having properties intermediate between those of glutenin and gliadin.

The mechanical spectra of gliadin and glutenin sub-fractions from good (cv. Hereward) and poor (cv. Riband) bread making wheat were compared to understand the relative importance of these groups of proteins in inter-cultivar variation in gluten viscoelasticity. The rheological parameters, G¢ and tan d, of gliadin fractions were similar for good and poor bread making wheats. However, there were significant differences in the G¢ and tan d values of the glutenin fractions from two wheats. This observation confirms that the inter-cultivar variation in gluten viscoelasticity is related to differences in the rheological properties of the glutenin fraction in the gluten.

The gliadin to glutenin ratio is another contributor to the variation in gluten viscoelasticity. Experiments, in which the gliadin / glutenin ratios were varied by replacing glutens with different levels of either gliadin or glutenin indicated an inverse relationship between the gliadin/glutenin ratio and the elasticity for both the good and poor glutens. The decrease in elasticity with increasing gliadin/

glutenin ratio may be attributed to a plasticising effect of gliadin and the 'interference' of gliadin with the glutenin-glutenin interactions. Interestingly, the elastic modulus, G¢, was more sensitive to changes in the gliadin/glutenin ratio at the lower end of the range examined (< about 1.0 to 1.5) than at the upper end.

The differences in elastic characters between the two glutens were observed to be less prominent at lower gliadin/glutenin ratios (<about 1.0 to 1.5) and more prominent at higher ratios, suggesting a greater weakening effect of a high proportion of gliadin (>50%). This effect appeared to be more pronounced in the case of the poor quality cv. Riband gluten than in the case of the good quality cv. Hereward gluten, perhaps suggesting inter-cultivar differences in the interactions between the two fractions.

Although it is well established that the high M_r glutenin subunits play an important part in determining bread making quality, the relationship of these subunits with gluten viscoelasticity has not been extensively explored. Recent studies demonstrated that the elastic modulus, G¢, classified the wheat cultivars into three groups on the basis of the subunits encoded at the *Glu A1* and *Glu-D1* loci (Fig. 4). Null alleles at the *Glu-A1* and subunits 2+12 at the *Glu-D1* locus were found to be related to poor elastic character. The presence of subunit 2* in cultivars containing subunits 2+12 or 3+12 increased elasticity substantially. Cultivars having subunits 2* or 1 at the *Glu-A1* and subunit pair 5+10 at the *Glu-D1* demonstrated the highest modulus of elasticity. According to some workers, the presence of subunits 2+12 instead of subunit pair 5+10 in a near-isogenic line resulted in a substantial decrease in gluten elasticity. Additionally, deletion of the subunits encoded

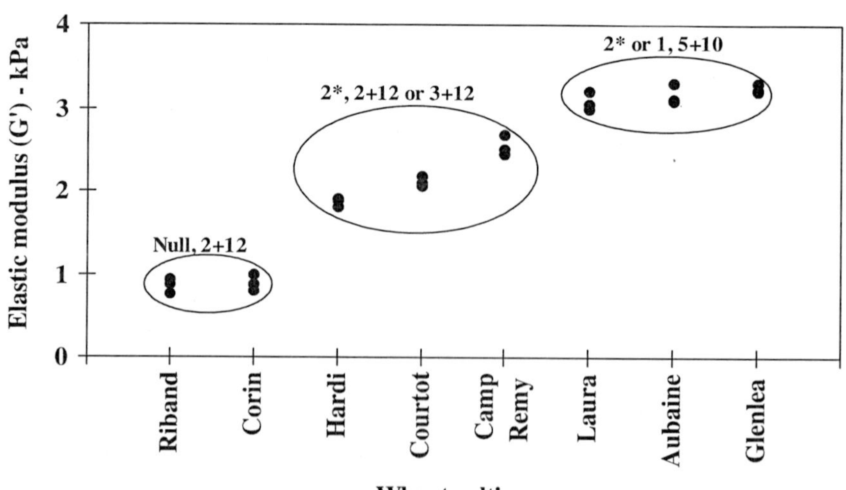

Fig. 4: Qualitative effects of high M_r glutenin subunits on the elastic modulus of the glutens

at the *Glu-A1* and *Glu-D1* loci caused a substantial decline in the elastic character of the gluten. Likewise, scientists have also found a significantly greater influence of subunits 5+10 on the elasticity of gluten than those of subunits 17+18. It was also noted that *Glu-1* triple deletion lead to a substantial decline in gluten viscoelasticity. In these investigations, the differential effects of high M_r glutenin subunits were attributed to the fact that the loss of subunits 5+10 in genetic lines caused a significantly greater reduction in the large polymers than those of subunits 2+12 or 17+18. This observation is consistent with the suggestion that subunit 5 of cv. Cheyenne contains an additional cysteine residue compared with subunit 2 of cv. Chinese spring, which could lead to the formation of a large, highly cross-linked polymer system conferring greater elastic character on the gluten network. However, the relative contribution of $1D_x5$ and $1D_y10$ subunits to quality associated with subunit pair 5+10 is still a matter of speculation.

The quantity of high M_r glutenin subunits is another important determinant of gluten viscoelasticity. In this work, the relative quantities of total high M_r glutenin subunits showed a second degree polynomial relationship with the G¢ values and loaf volume data of glutens, which explained 74% and 73% of the variation in the elastic character and bread making quality of the glutens.

Similarly, the high to low M_r glutenin subunit ratio showed a significant ($R^2 = 0.60$; $P < 0.009$) direct relationship with gluten elastic character, again emphasizing the relative importance of high M_r subunits for gluten viscoelasticity (results not shown). It is speculated that the enhanced elasticity of gluten caused by increasing the proportion of the high M_r subunits of glutenin might be linked to the fact that they have been proposed as forming the backbone of the glutenin network. Therefore, increasing the concentration of these subunits might provide conditions conducive to extensive cross-linking. The type of subunits, indeed, may also enhance cross-linking, thus, in effect, raising the average molecular size of the polymer, and hence the elasticity and bread making quality.

Finally, one ultimate objective of any rheological test is to determine the baking potential of wheat cultivars without performing a baking test, which would help to save both time and money. Here controlled stress rheometry was tried to assess the bread making performances of isolated glutens. For this purpose, loaf volume data were obtained with reconstituted flours, in which equal amounts of gluten protein from each cultivar were added to starch and water solubles obtained from cv. Hereward. Unlike flour-water doughs, the G¢ values of gluten-water doughs showed a significant direct relationship with bread making performance, explaining 73% of the variation in loaf volume (Fig. 5). So clearly, fundamental rheological tests on gluten can be used to predict the bread making qualities of wheat cultivars.

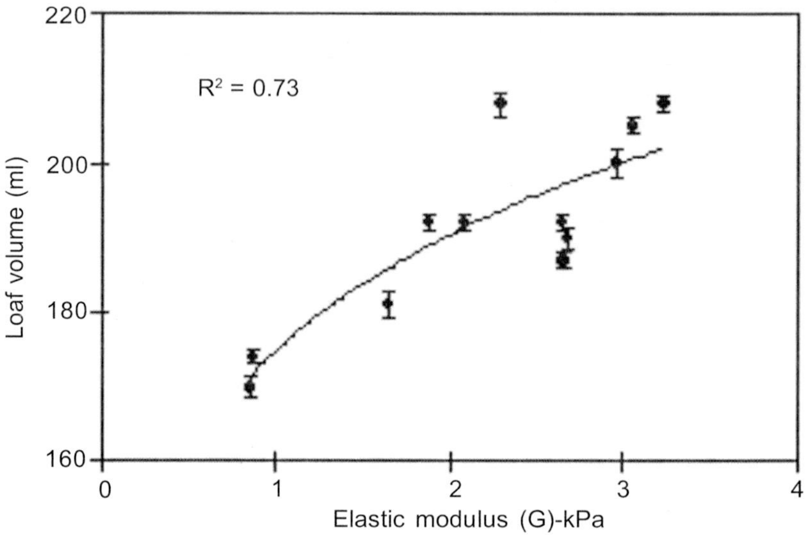

Fig. 5: The relationship between loaf volume (reconstituted flour) and elastic modulus of glutens from different wheat cultivars

CONCLUSIONS AND FUTURE RESEARCH NEEDS

The technological significance of wheat flour is related to the physicochemical and structural properties of wheat gluten proteins. In this context, particular attention has been focused upon the glutenin fraction of wheat gluten since it is qualitative and quantitative variation in this fraction that accounts for a substantial proportion of the variation in gluten viscoelasticity and breadmaking quality that occurs amongst wheat cultivars. In the past genetic lines have proved to be powerful tools in investigating the effect of specific high M_r subunits on wheat dough and gluten functionality. Although it is now evident that the high M_r subunits of glutenin play a significant role in gluten viscoelasticity and breadmaking performance, the basis of the differential effect of various glutenin polypeptides on functionality of wheat flour remains to be defined. Also, little is known concerning the role of low M_r glutenin subunits and gliadin subgroups in gluten viscoelasticity and breadmaking potential. Therefore, further research is needed in this area as well.

9

Emerging Technology in Quality Evaluation

S. N. Jha

Abstract

Consumers are now more conscious about quality and source of their foods. Attempts made to determine the quality of food materials are numerous, but most of them are destructive in nature. In recent years, nondestructive methods of quality evaluation have gained momentum and a considerable attempts have been made to develop them. This paper describes application of these non-emerging technologies: magnetic resonance imaging, x-ray computer tomography, near infrared spectroscopy and a few other important techniques for determination of different quality attributes with special emphasis on grain, and discusses their pros and cons for practical exploitation.

INTRODUCTION

The ultimate objective of the production, handling and distribution of food materials including grain is to satisfy consumers. It is generally agreed that consumer satisfaction is related to product quality. While the term quality has been defined in many ways and contexts, there is little agreement as to what it is, how it can be measured, and how it relates to consumer acceptability. Quality can be viewed as an absence of defects or a degree of excellence. Quality means different things to different handlers within the distribution chain. Food quality embraces both sensory attributes that are readily perceived by the human senses and hidden attributes such as safety and nutrition that require sophisticated instrumentation to measure. Quality of a fresh fruit or vegetable changes as it proceeds from harvest to consumer. Similar is the case with grain stored for future consumption and supply. The relative importance of different quality attributes changes from handling to purchase to consumption. What quality is then depends on the perspective of the viewer. An understanding of the different perspectives of different participants in post harvest distribution is essential

Product quality and quality evaluation methods are naturally extremely important. The decisions concerning harvesting, maturity, ripeness, and quality are based mostly on subjective and visual inspection of external appearance. Several non-destructive techniques for quality evaluation have been developed based on the detection of various physical properties that correlate well with certain factors of a product. The quality of grains is mostly based on size, shape, color, gloss, flavor, firmness, texture, taste, freedom from foreign matters, and external as well as internal defects. Numerous emerging techniques for evaluating the above external quality factors are now available commercially. Internal quality factors such as maturity, sugar content, acidity, oil content, protein and vitamins contents, and internal defects, however, are difficult to evaluate.

Methods are needed to better predict the internal quality without destroying the sample and using the techniques on line and on the spot. Recently, there has been as increasing interest in nondestructive methods of quality evaluation, and a considerable amount of effort has been made in that direction. But the real problem is how these methods are to be exploited practically and what the difficulties are in implementing them. The objective of the present paper is thus to discuss the application of the most recent nondestructive methods such as nuclear magnetic resonance, x-ray computed tomography, near-infrared spectroscopy and some other important methods and to evaluate their pros and cons for suitability in commercial application.

Nuclear magnetic resonance (NMR) techniques

The nuclear magnetic resonance technique, often referred as magnetic resonance imaging (MRI), involves resonant magnetic energy absorption by nuclei placed in an alternating magnetic field. The amount of energy absorbed by the nuclei is directly proportional to the number of a particular nucleus in the sample such as the protons in water oil. The theory of NMR is presented in detail elsewhere (Farrar & Becker, 1971). The basic concepts, types of pulsed experiments and the type of information that can be extracted from these experiments are described. Information on experimentation, assembling hardware, conducting laboratory tests and interpreting the results is also available from Fukushima and Roeder (1981). These authors also provided detailed theory for better understanding of what a scientist should seek and what he might expect to find out by using NMR.

There are many applications of NMR in agriculture (Rollwitz, 1984). The simplest among them is the determination of moisture and oil content (Mousseri *et al.*,

1974, Leung *et al.*, 1976; Miller *et al.*, 1980; Brosio *et al.*, 1978; Rollwitz and Persyn, 1971). But the NMR response many times is not clear and poses problems especially when constituents other than water are present in the material (Steinberg & Richardson, 1996). Besides the established relationship between the moisture and output of NMR experiments, various other facts helpful in determining the quality of food materials without destroying them are available in the literature: Selections of chocolate confectionary products can be made non-invasively by three-dimensional magnetic resonance imaging (Miquel *et al.*, 1998); using a spin echo pulse sequence, 128x64x64 data sets were acquired with either a 5-or 20-ms echo time, 500-ms repetition time and signal averages, in total 2 h scan time. Such images localize and distinguish between the constituents, and visualize both the internal and external structure of matter.

Most perishable food products are now marketed in packaged form. To increase the marketability longer shelf life is needed and this is achieved by freezing and secondary processing of the food. During freezing it is natural that ice will form within the food that may change its characteristics. Ice formation during food freezing-can be examined using the NMRI method as the formation of ice has been seen to reduce the spatially located NMR signal. The characteristics of a food can be better controlled as MRI can serve to assess freezing times and the food structure during the freezing process (Kerr *et al.*, 1998). The secondary processing changes almost all characteristics of a food, such as physical and aerodynamic (Jha & Kachru, 1998), thermal and hygroscopic properties (Jha & Prasad, 1993; Jha 1999), which in turn, change it key acceptability factors, i.e. sensory texture and taste. The sensory texture of cooked food such as potatoes has been predicted using the NMRI technique (Thybu *et al.*, 2000). In addition, NMR image intensity, the ratio of the oil and water resonance peaks of the one-dimensional NMR spectrum, and both the spin-lattice relaxation time and spin-spin relaxation time of water in the fruit are correlated with maturity of a fruit like avocado before harvesting (Chen *et al.*, 1993). This important finding has desirable features for high speed sorting using a surface-coil NMR probe that determine the oil/water resonance peak ratio of the signal from one region in an intact fruit.

An on-line nuclear magnetic resonance quality evaluation sensor has recently been designed, constructed, and tested (Kim *et al.*, 1999). The device consists of a super-conducting magnet with a 20mm diameter surface coil and a 150 mm diameter imaging coil coupled to a conveyor system. These spectra were used to measure the oil/water ratio in avocados and this ratio correlated to percent dry weight. One-dimensional magnetic resonance images of cherries were later used to detect the presence of pits inside.

X-ray and computerized tomography (CT)

X-ray imaging is an established technique to detect strongly attenuating materials and has been applied to a number of inspection applications within the agricultural and food industries. In particular, there are many applications within the biological sciences where we wish to detect weakly attenuating materials against similar background material.

X-ray computed tomography (CT) has been used to image interior regions of apples with varying moisture and, to a limited extent, density states (Tollner *et al.*, 1992). The images were actually maps of x-ray absorption of fruit cross sections. X-ray absorption properties were evaluated using normal apples alternatively canned and sequentially freeze-dried, fruit affected by water core disorder, and normal apples freeze-dried to varying levels. The results suggested that internal differences in x-ray absorption within scans of fruit cross-sections are largely associated with differences in volumetric water content. Similarly, the physiological constituents have been monitored in peaches by CT methods in which x-ray absorbed by the peaches is expressed in CT number and used as an index for measuring the changes in internal quality of the fruit (Barcelon *et al.*, 1999). Relationships between the CT number and the physiological contents were determined and it was concluded that x-ray CT imaging could be an effective tool in the evaluation of peach internal quality.

In another study, the potential for Compton scattered x-rays in food inspection was evaluated by imaging the density variation across a food material by measuring the Compton scatter profile across a food material by measuring the Compton scatter profile across polystyenespheres with internal voids (MacFarlane *et al.*, 2000). In this study particular attention was paid to simulate the obscuring influence of multiple scatter. The simulated result was found to be in close agreement with the experimental observation. Some experimental test sample of a Perspex block with various embedded soft materials showed that care should be taken to ensure that the transmission image is taken with x-ray within an appropriate energy range (Zwiggelaar *et al.*, 1997). For low Z materials the contrasts between the materials became more pronounced at lower x-ray energies. If more than one soft material has to be distinguished from the surrounding area it may be advantageous to image over a range of x-ray energies.

Near-infrared spectroscopy

The use of near-infrared spectroscopy as rapid and often nondestructive technique for measuring the composition of biological materials has been demonstrated for many commodities. This methods is no longer new; as it started in early 1970 in Japan (Kawano, 1998), Just after some reports from America. Even an official method to determine the protein content of wheat is available (AACC, 1983). The National Food Research Institute (NFRI), Tsukuba has since become a leading institute in NIR research in Japan and has played a pivotal role in expanding near-infrared spectroscopy technology all over the country (Iwamoto *et al.*, 1995). In Japan, NIR as a nondestructive method for quality evaluation was started for the determination of sugar content in intact peaches, Satsuma orange and similar other soluble solids (Kawano, 1994).

To determine the solid content of cantaloupe Dull *et al.* (1989) used NIR light at 884 nm and 913 nm. Initially the correlation of their findings was poor mainly due to light losses. Later, Dull and Birth (1989) modified the earlier method and applied it to honey-dew melons; the improved methods showed better correlation. Similarly, a nondestructive optical method for determining the internal quality of intact peaches and nectarines was investigated (Slaughter, 1995). Based upon visible and near-infrared spectrophotometer techniques, the method was capable of simultaneously predicting the soluble solid content, sucrose content, sorbitol content, etc. of intact peaches and nectarines was investigated (Slaughter, 1995). Based upon visible and near-infrared spectrophotometer techniques, the method was capable of simultaneously predicting the soluble solid content, sucrose content, sorbitol content, etc. of intact peaches and nectarines, and required no sample preparation.

Now various NIR spectrometers are available and are being used commercially. Some modifications in these available spectrometers, especially for holding the intact samples, are reported (Kawano *et al.*, 1992; 1993). In the same sample holding a test tube for holding liquid food such as milk was also used to determine fat content (Chen *et al.*, 1999). Recently a low cost NIR spectrometer has been used to estimate the soluble solids and dry matter content of kiwifrui (Osborne & Kunnemeyer, 1999). Errors are within the permissible limit and the time requires for obtaining data has been reduced. The influence of sample temperature on the NIR calibration equation was also evaluated and a compensation curve for the sample temperature was developed (Kawano *et al.*, 1995) to rectify the result.

Miscellaneous techniques

Quality attributes such as invisible surface bruises, color, gloss, firmness, density, volume expansion of processed food etc. are also important (Jha & Prasad, 1996). Often consumers select food materials, particularly fruits and vegetables by judging these parameters visually. Multiple efforts have been made to determine these parameters visually. A fluorescence technique was used to detect invisible surface bruises on Satsuma mandarins (Uozumi *et al.*, 1987). The authors have also tested this method successfully to know the freshness of cucumbers and eggs and found it very useful for detecting the freshness of agricultural produce.

Matsuoka *et al.*, (1995) measured the gloss of eggplant by a spectral radiometer system and found or to be a viable parameter for determining freshness. They observed remarkable change in relative spectral reflectance values after 48 h. Later, they compares their evaluation by eye in a sorting house with the integrated results of relative spectral reflectance in the visible range and found that the gloss on the surface differs with light and is caused by round and adhesives substances on the epidermal cells (Matsuoka *et al.*, 1996). A unique gloss meter for measuring the gloss on the surface differs with light and is caused by round and adhesives substances on the epidermal cells (Matsuoka *et al*., 1996). A unique gloss meter for measuring the gloss of curved surfaces was used in parallel with a conventional, flat surface gloss meter to measure peel gloss of ripening banana (Ward & Nussinovitch, 1996). Usually banana ripeness is judged by the color of the peel. The new gloss meter is able to measure the peel correctly which helps in predicting the correct time and level of ripening. This is also able to measure the gloss of other fruits and vegetables such as green bell pepper, orange, tomato, eggplant, and onion (Nussinovitch *et al*., 1996).

Glossiness and color, in fact, are the only visual attributes for measuring the quality of fruits and vegetables. Another property that helps a consumer in deciding the quality is firmness. Takao (1998) developed a fruit hardness tester that can measure the firmness of kiwifruit nondestructively. The tester is called a 'HIT counter' after the three words, hardness, immaturity and texture. By just setting the sample in the tester, the amount of change in shape is measured and a digital reading within a few seconds indicates about the freshness. Based on the same principal another on-line prototype HIT counter, fruit hardness sorting machine has also been developed (Takao & Omori, 1991). The relationship between density and internal quality of watermelon can also be determined. An optimum range of density was first determined and then a new automatic density sorting system was develops and then a new automatic density sorting system was developed to measure the hollowness of a watermelons with cavities or deteriorated porous flesh to be removed and permits estimation of the soluble solid content of this fruits.

Neural networks have lately gained in popularity as an alternative to regression models to regression models to characterize the biological process. Their decision-making capabilities can best to used in image analysis of biological products where the shape and size classification is not governed by any mathematical function. Many neural network classifiers have been used and evaluated for classifying agricultural products, but multi-layer neural network classifiers perform such tasks best (Jayas *et al*., 2000). Recently one scientist used a gamma- absorption techniques combined with a scanning device for continuous non-destructive crop mass and growth measurement in the field (Gutezeit, 200). Most important in this study was the accuracy of the measurement, which was found to be in agreement with the direct weighting system. This method has made it possible to assess the reaction of plants and their dependence on environmental factors by growth analysis.

CONCLUSIONS

Determination of quality of any food material is actually a complex problem that requires a variety of specific sensor, more than an accumulation of simple sensor. Various techniques are being tried. IMR, x-ray CT and NIR techniques may be useful for a large volume of work in agriculture, especially for evaluation of qualities such as maturity, internal quality of fruit and conditions of food materials after processing. These techniques, although give a correct picture and precise measurement of parameters, are not convenient for small business. Their high cost restricts application to large entrepreneurs and developed countries only.

Two examples of the use of x-ray imaging relevant to the agricultural and food industries have been given, notably in the inspection of vegetables and food materials using low energy x-ray imaging and in the inspection and control of dynamic processes. The x-ray imaging results have been compared with the full three-dimensional information obtained by computer tomogrphy. The CT results show more detail in the test sample than the single transmission image and detail in the inspection of materials of variable shape usually encountered in the agricultural and food industry. The imaging techniques MRI and x-ray CT are able to show only the internal structure of the material, not the compositional of nutritional details, whereas NIR techniques are very successfully being used to determine the compositional quality of a food. However, it is not yet possible to produce an image of the internal physical quality of food materials. All techniques are costly and their handling requires scientific knowledge. Research must continue to focus on making their use easier and lowering the cost to make it within the reach of common business/growers of both developed and developing countries.

REFERENCES

American Association of Cereal Chemists. (1983). Approved methods of the Am. Assoc. of Cereal Chemists, 8th Ed. (March). St. Paul, Minn.

Barcelon, E.G., Tojo, S. and Watanable, K. (1999). X-ray computed tomography for internal quality evaluation of peaches. J. Agric. Eng. Res., 73, 323-330.

Brosia , E.F., Conti, C.L. and Sykara, S. (1978). Moisture determination in starch-rich food products by pulsed nuclear magnetic resonance. J. Food Technol., 13, 107-116.

Chen, J.Y., Iyo, C. and Kawano, S. (1999). Development of calibration with sample cell compensation for determining the fat content of unhomogenised raw milk by a simple near infrared transmittance method. J. Near Infrared Spectosc., 7, 265-273.

Chen, P., McCarthy, M.J., Kauten E., Sarig, Y. and Han, S. (1993). Maturity evaluation of avocados by NMR methods. J. Agric. Eng. Res., 55, 177-187.

Dull, G.G. and Birth, G.S. (1989). Nondestructive evaluation of fruit quality: Use of NIR spectrophotometry to measure solube solids in intact honeydew melons. Hortsci. 24, 754.

Dull, G.G. and Birth, G.S. Smittle, D.A and Leffler, R.G. (1989). Near infrared analysis of soluble solids in intact cantaloupe. J. Food Sci., 54, 393-395.

Farrar, T.C. and Becker, E.D. (1971). Pulse and Fourier Transform NMR: Introduction to Theory and Methods. Academic Press, New York.

Fukushima, E. and Roeder, S.B.W. (1981). Experimental Pulse NMR. Addison-Wesley Publishing Company, Reading, M.A.

Gutezeit, B. (2000). Non-destructive measurement of fresh plant mass by the gamma-scanning technique applied to broccoli. J. Agric. Eng. Res., 75, 251-255.

Iwamoto, M., Kawano, S. and Yukihiro, O. (1995). An overview of research and development of NIR spectroscopy in Japan. J. Near Infrared Spectrosc., 3: 179-189.

Jayas, D.S., Paliwal, J and Visen, N.S. (2000). Multi-layer neural networks for image analysis of agricultural products. J. Agric. Eng. Res. (in press).

Jha, S.N. (1999). Physical and hygroscopic properties of makhana. J. Agric. Eng. Res., 72, 145-150.

Jha. S. N. and Kachru, R.P. (1998). Physical and aerodynamic properties of makhana. J.Food Process.Eng., 79, 301-316.

Jha, S.N. Matsuoka T. (2000). Review: Nondestructive Techniques for quality evaluation of intact fruits and vegetables. Food Sci and Tech Res, 6(4), 248–251.

Jha, S.N. and Prasad, S. (1993). Physical and thermal properties of gorgon nut. J. Food Process Eng., 16, 237-245.

Jha, S.N. and Prasad, S. (1996). Determination of processing conditions of gorgon nut (*Euryale ferox*). J. Agric. Engg. Res., 63, 103-112.

Kato, K. (1997). Electrical density sorting and estimation of soluble solids contents of watermelon. J. Agric. Engg. Res., 67, 161-170.

Kawano, S. (1994). Nondestructive near infrared quality evaluation of fruits and vegetables in Japan. NIR News, 5, 10-12.

Kawano, S. (1998). New application of nondestructive methods for quality evaluation of fruits and vegetables in Japan. J. Jpn. Soc. Hort. Sci., 67, 1176-1179.

Kawano, S., Abe, H. and Iwamoto, M. (1995). Development of a calibration equation with temperature compensation for determining the brix value in intact peaches. J. Near infrared Spectrosc., 3, 211-218.

Kawano, S., Fujiwara, T., and Iwamoto, M.C. (1993). Nondestructive determination of sugar content in satsuma maddarin using NIR transmittance. J. Jpn. Soc. Hort. Sci., 62: 465-470.

Kawano, S., Watanabe, H. and Iwamoto, M. (1992). Determination of sugar content in intact peaches by NIR spectroscopy with fibre optics in interactance mode. *J. Jpn. Soc. Hort. Sci.,* 61, 445-451.

Kerr, W.L., Kauten, R.J., McCarthy, M.J. and Reid, D.S. (1998). Monitoring the formation of ice during food freezing by NMR sensors. J.Agri.Eng. Res., 74, 293-301.

Kim, S.M., Chen, P., McCarthy, M.J. and Zoin, B. (1999). Fruit internal quality evaluation using on-line NMR sensors. J. Agric. Eng. Res., 74, 293-301.

Leung, H.K., Steinberg, M.P. Wei, L.S. and Nelson, A.I. (1976). Water binding of macromolecules determined by pulsed NMR. J. Food Sci., 41, 297-300.

Macfarlane, N.J.B., Bull, C.R., Tillett, R.D., Speller, R.D., Royle, G.J and Johnson, K.R.A. (2000). The potential for compton scattered x-rays in food inspection: The effect of multiple scatter and sample inhomogeneity. J. Agric. Eng. Res., 75, 265-274.

Matsuoka, T., Miyauchi, K. and Sun, D. (1995). Basic studies on the quality evaluation of agricultural products-Quantification of gloss of eggplants by spectral characteristics. J. Jpn. Sci. Agric March., 57, 33-40.

Matsuoka, T., Miyauchi, K. and Yano, T. (1996). Basic studies on the quality evaluation of agricultural products (Part 2)- The evaluation of colour and gloss decrease on the surface of eggplants. J. Jpn. Sci. Agric. Mach., 58, 69-77.

Miller, B.S., Lee, M.S., Hughes, J.W. and Pomernz, Y. (1980). Measuring high moisture content of cereal grains by pulsed nuclear magnetic resonance. Cereal Chem. 57, 126-129.

Miquel, M.E., Evans, S.D. and Hall.L.D. (1998). Three-dimensional imaging of chocolate confectionary by magnetic resonance methods. J.Fd Sci.Tech., 31, 339-343.

Mousseri, J., Steinberg, M.P., Nelson, A.I. and Wei, L.S. (1974). Bound water capacity of corn starch and its derivatives by NMR. J. Food Sci., 39, 114-116.

Nussinovitch, A., Ward, G. and Mey-tal, E. (1996). Gloss of fruits and vegetables. J. Food Sci. Technol., 29, 184-186.

Osborne, S.D. and Kunnemeyer, R. (1999). A low cost system for the grading of kiwifruit. J. Near Infrared Spectocs., 7, 9-15.

Rollwitz, W.L. (1984). Radio frequency spectroscopy: Versatile sensors for agriculture applications. ASAE Paper No. 84-1590, ASAE. St. Joseph, MI 49085.

Rollwitz, W.L. and Persyn, G.A. (1971). On-stream NMR Measurements and control. Am. Oil Chem. Soc. J., 48, 59-66.

Slaughter, D.C. (1995). Nondestructive determination of internal quality in peaches and nectarines. Trans. ASAE, 38, 617-623.

Steinberg, M.P. and Richardson, S.J. (1986). Application of nuclear magnetic resonance. Abstract Volume: International Union Food Science and Technology Symposium on Water Activity: Theory and Application. June 13-14, Dallas, TX. 11.

Takao, H. (1988). HIT counter. Noryu Giken Kaihou, 180, 7-9 (in Japanese).

Takao, H. and Omori, S. (1991). Quality evaluation of fruits and vegetables using light transmittance. Noryu Giken Kaihou, 145, 14-16 (in Japanese).

Thybo, A.K., Bechmann, I.E., Martens, M. and Engelsen, S.B. (200). Prediction of sensory texture of cooked potatoes using uniaxial compression, near infrared spectroscopy and low field H NMR Spectroscopy. Food Sci. Technol., 33, 103-111.

Tollner, E.W., Hung, Y.C., Upchurch, B.L. and Prussia, S.E. (1992). Relating x-ray absorption to density and water content in apples. Trans. ASAE, 35, 1921-1928.

Uozumi, J., Kawano, S., Iwamoto, M. and Nishnari, K. (1987). Spectrometric system for the quality evaluation of unevenly coloured food. Nippon Shokuhin Kagyo Gakkaishi, 34, 163-170 (in Japanese).

Ward, G. and Nussinovitch, A. (1996). Peel as an indicator for banana ripeness. Food Sci. Techol., 29, 289-294.

Zwiggelaar, R., Bull, C.R., Mooney, M.J. and Czarnes, S. (1997). The detection of soft materials by selective energy x-ray transmission imaging and computer tomography. J. Agric. Eng. Res., 66, 203-21.

10

Water Absorption Characteristics of Green Gram and Black Gram

Monesh, M. Pethe, Jagdish, S. Nikhade, V.B. Kalmegh

Abstract

Pulses being nutritionally vital, is an important constituent of diet for a very large number of peoples. A part from being a relatively cheaper source of dietary protein, it is also an important source of calories, certain essential amino acid, minerals and vitamins. Post harvest scenario of pulse shows that 75 per cent of production was milled in pulses with as dehusked split i.e. dal. Traditionally green gram and black gram dal is dehulled manually after soaking in water. Present investigation was carried out to develop the wet type dehuller and study was carried out on water absorption characteristics of green gram and black gram. The result reveals that the dehulling efficiency of abrasion type mechanism was about 95 per cent and 92 per cent in case of shearing type mechanism. It has been observed that rate of water absorption was initially high and then gradually decreases i.e. 0.54 to 0.18 and 0.48 to 0.18 for black and green gram, respectively.

INTRODUCTION

Green gram (*Vigna radiata*) and black gram (V*igna mungo*) are the pulses with annual production of 1.29 million tonnes and 1.31 million tonnes respectively. The area under green gram and black gram against world in country is 3.13 million hectares and 3.18 million hectares respectively. The per capita availability of pulses in India is currently estimated as 24 g/day as moderately recommended intake 42 g/day (*Indian Agricultural Survey, Hindu, 1998*).

Pulses, the dry edible seeds of various leguminous plants are used for human food throughout the world and India is not an exception to it. Pulses contain 20-25 per cent protein and are a good source of protein for predominantly vegetarian populations and others who depend on plant proteins only.

Pulses being nutritionally vital are an important constituent of diet for a very large number of peoples. A part from being a relatively cheaper source of dietary protein, it is also an important source of calories, certain essential amino acid, minerals and vitamins. Post harvest scenario of pulses shows that 75 per cent of production was milled in pulses with as dehusked splits i.e. dal. Traditionally green gram a black gram is dehulled manually after soaking in water. The efficient and improved dehulling of pulses is of vital importance in reducing dehulling losses and thus increasing availability of dal in daily diet of people. Dehusking reduces the cooking time and antinutrients, it also improves palatability and digestibility for that a study was taken to develop the wet type dehuller and evaluate the water absorption characteristics on green gram and black gram.

MATERIALS AND METHODS

The samples (5 g each) of green gram and black gram were soaked in 50 ml of water for soaking period of 5 hours. At an interval of one hour the sample was drained surface moisture was removed and weightments were carried out. A process flow chart is shown in Fig.1.

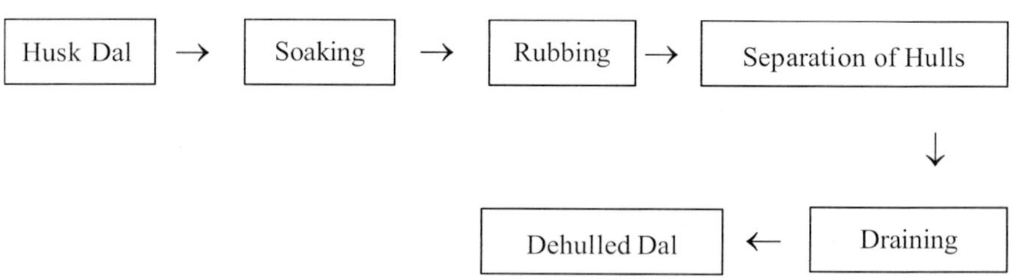

Fig.1: Process flow chart for dehulling of Pulses (Dal)

Performance of dehulling was evaluated by two different mechanism were developed for dehulling of soaked green gram and black gram.

1. Shearing type mechanism
2. Abrasion type mechanism

The water absorption (%) and rate of water absorption (g/h-g) was determined by given formula: Water absorption (%) $= (W_2 - W_1 / W_1) \times 100$

Rate of water absorption (%) = $(W_2 - W_1) / (W_1 \times \theta)$

Where, W_1 - initial weight of sample prior to soaking, g; W_2 - final weight of sample after soaking, g; θ - soaking period, h

RESULTS AND DISCUSSION

It was observed that the rate of water absorption varied from 0.48 to 0.17 (g/h-g) in green gram and 0.52 to 0.19 (g/h-g) in black gram during soaking period of 5 hrs as shown in Table 1.

Table 1: Effect of ambient water soaking (30 °C) on absorption characteristics

Soaking Time, h	Moisture content, %	Sample weight, g	Water absorption,%	Rate of water absorption, g/h-g
Green gram				
0.	10.91	5.00	-	-
1.	99.00	7.40	48.00	0.48
2.	146.00	8.80	76.00	0.38
3.	167.00	9.20	84.00	0.28
4.	171.00	9.28	85.60	0.22
5.	173.00	9.30	86.00	0.17
Black gram				
0.	10.62	5.00	-	-
1.	84.00	7.60	52.00	0.52
2.	135.00	8.80	76.40	0.38
3.	150.00	9.45	89.00	0.30
4.	156.00	9.60	92.00	0.23
5.	159.00	9.70	94.00	0.19

During soaking period of 5 hours rate of water absorption decreased as 0.16, 0.17 and 0.19 (g/h-g) in green gram and 0.18, 0.19 and 0.2 (g/h-g) in black gram with respect to 35, 40 and 45°C temperature is given in Table 2.

Table 2: Effect of warm water soaking on absorption characteristics

Soaking Time, h	Water absorption, %			Rate of water absorption, g/h-g		
	35°C	40°C	45°C	35°C	40°C	45°C
Green gram						
1.	60.00	64.00	80.00	0.60	0.64	0.80
2.	70.00	70.00	84.40	0.35	0.35	0.42
3.	74.00	78.00	87.00	0.25	0.26	0.29
4.	80.00	86.00	93.00	0.20	0.21	0.23
5.	82.00	90.00	96.00	0.16	0.18	0.11

(Contd.)

Soaking Time, h	Water absorption, %			Rate of water absorption, g/h-g		
	35°C	40°C	45°C	35°C	40°C	45°C
Black gram						
1.	62.00	68.00	78.00	0.62	0.68	0.78
2.	74.00	74.00	85.00	0.37	0.37	0.42
3.	78.00	80.00	90.00	0.26	0.30	0.30
4.	82.00	90.00	96.00	0.20	0.23	0.24
5.	84.00	96.00	98.00	0.18	0.19	0.20

As the temperature increases rate of water absorption decreases. The average value of physical dimensions of green gram and black gram increase to 6.80, 4.41 and 2.31 mm and 6.94, 4.55 and 2.65 mm respectively as shown in Table 3.

Table 3: Effect of water absorption on linear dimensions

Soaking time, h	Moisture content, %	Dimensions, mm			Sphericity	Volume, mm^3
		l	b	t		
Green gram						
0	10.91	4.90	3.16	1.85	0.62	25.62
1	99.00	5.85	4.32	2.00	0.62	57.26
2	146.00	6.48	4.38	2.13	0.62	64.49
3	167.00	6.52	4.40	2.27	0.62	66.09
4	171.00	6.80	4.41	2.36	0.61	69.59
5	173.00	6.80	4.41	2.37	0.61	69.59
Black gram						
0	10.62	5.20	3.91	2.19	0.66	16.50
1	84.00	6.10	3.40	2.59	0.66	61.63
2	135.00	6.40	4.46	2.53	0.65	66.25
3	150.00	6.93	4.50	2.60	0.63	73.37
4	156.00	6.94	4.55	2.65	0.63	75.50
5	159.00	6.94	4.55	2.65	0.63	76.89

The maximum dehulling efficiency (95.5 %) and (94.05%) in black gram of abrasion type dehuller and minimum (91.82%) and (92.39%) in green gram of shearing type is given in Table 4 and Table 5, respectively.

Table 4: Dehulling performance of shearing type dehuller

S. No.	Weight of different fractions, g					Dehulling efficiency, %
	Dehulled beans	Unhulled beans hulls + cotyledons	Broken beans	Hulls	Total	
Green Gram						
1	83.19	0.84+2.37	1.83	9.55	98.33	91.92
2	83.21	0.86+2.86	1.85	9.56	98.34	91.95
3	83.23	0.88+2.83	1.85	9.6	98.4	91.6
Average	83.21	0.86+2.68	1.84	9.57	98.35	91.82
Black gram						
1	83.10	0.80+2.56	1.44	10.54	98.50	92.46
2	83.15	0.86+2.57	1.43	10.56	98.59	92.31
3	83.18	0.87+2.59	1.40	10.58	98.69	92.4
Average	83.14	0.84+2.57	1.42	10.56	98.59	92.39

Table 5: Dehulling performance of abrasion type dehuller

S. No.	Weight of different fractions, g					Dehulling efficiency, %
	Dehulled beans	Unhulled beans hulls + cotyledons	Broken beans	Hulls	Total	
Green Gram						
1.	83.60	0.43+1.33	2.27	9.33	96.96	95.60
2.	83.80	0.45+1.30	2.25	9.35	97.15	95.71
3.	83.90	0.46+1.28	2.22	9.37	97.25	95.21
Average	83.77	0.45+1.30	2.25	9.35	97.12	95.50
Black gram						
1.	83.33	0.65+1.74	2.00	10.27	98.00	94.65
2.	83.34	0.67+1.72	2.00	10.28	98.00	93.86
3.	83.50	0.70+1.70	1.99	10.30	98.04	93.64
Average	83.34	0.67+1.72	1.99	10.28	98.02	94.05

CONCLUSIONS

A dehuller was developed for green gram and black gram at CAET, Akola. The dehulling efficiency of abrasion type mechanism was about 95 per cent and 92 per cent in case of shearing type mechanism. It was concluded that the water absorption rate was higher in case of black gram splits (0.19 g/h-g) as compared to green gram (0.17 g/h-g) and increases with the increase in temperature. It also concluded that the rate of water absorption was decreased wit increase in soaking period. The volume increases by 2.71 time in green gram, while 4.66 times in black gram and soaking time increase, linear dimensions of green gram and black gram also increases.

REFERENCES

Anonymous, (1998). Indian Agricultural Survey. The Hindu1998.

Agrawal, U.S. (2000). Comparison of various pre-milling treatments of pegeon pea grain on a laboratory dhal mill. J.I.E. (India), 80:55-57.

Kulkarni, S.D. (1989). Premilling aspects of pulses processing - a review. J.I.E. (India), 69:73-76.

Mohsenin, N.N. (1984). Physical properties of plant and animal materials. Cardon and Breach Sci. Pub. New York (2nd Edition).

Patil, R.T.; Singh, J. and Bargale, P.C. (1991). Development of wet type dehuller for blanched soybean. J. Food Sci. Tech., 28(4): 234-236.

Phirke, P.S. and Bhole, N.G. (1997). Dehulling of pigeon pea grain with flat plate. J. I. E. (India), 78:1-6.

Sahay, K.M. and Singh, K.K. (1994). Unit operation of agricultural processing. Vikas Pub. House Pvt. Ltd. (India).

SECTION - III
VALUE ADDITION

11

Use of Emerging Technologies in Evaluating Wheat Genotypes for their Suitability for Value Addition Qualities

J.P. Jaiswal, Sohan Pal, D.P. Saini and Archana Vaishnava

Abstract

The constant improvement in automation of bakery industry has placed higher demands on wheat. As a consequence bakery industry requires wheat possessing specific quality attributes. In view of this, there is need to produce good quality wheat not only for good chapattis but also for better bakery products viz., bread, biscuit, noodles, pasta products. New technologies are emerging which could be exploited for determining the Indian wheats for their suitability for various value added products. In the current investigation, 50 genotypes possessing low protein content (<10.5%) selected from a vast amount of breeding material were analysed for their biscuit making attributes. These lines were screened by actual baking test and alkaline water retention capacity (AWRC) test keeping the standard prescribed for biscuit making qualities viz., low protein content, low sedimentation value, low AWRC and larger diameter with larger spread factor of biscuit. Further with the help of Stable Micro Systems TA.HDi Texture Analyser the biscuits were analysed for hardness and crispness. Correlation coefficient among 8 quality traits suggested highly significant and positive correlation between sedimentation value and AWRC, biscuit diameter and spread factor, and biscuit hardness and fracturability, and significant negative correlation between biscuit diameter and its thickness. On the basis of findings, genotypes identified for biscuit making qualities may be used as donor for developing varieties suitable for biscuit making.

INTRODUCTION

The global significance of wheat could simply be realized in the way that more foods are made with wheat than any other cereal. Wheat contributes between 10-12 per cent of the daily calorie intake in people over 60 countries at global level. The wheat trade and the value added products accounts for several billion US dollars. Having made an unprecedented growth in the production and the productivity levels of wheat in India, the focus of the research needs to be oriented towards quality breeding in a systematic manner. Our share in global market is negligible despite considerable wheat surplus. The constant improvement in automation of bakery industry has placed higher demands on wheat. As a consequence bakery industry requires wheat possessing specific quality attributes. In view of this, there is need to produce good quality wheat not only for good chapattis but also for better bakery products viz., bread, biscuit, noodles, pasta products and of course chapattis. New technologies are emerging which could be exploited for determining the Indian wheat for their suitability for various value added products. Identification of wheat genotypes for specific products would help in fetching better prices of their product as well as earning foreign exchange through export.

Commonly used tests for predicting wheat for biscuit making qualities are protein content, sedimentation value (for gluten strength), sugar-snap biscuit diameter and alkaline water retention capacity (AWRC). However, two more traits, biscuit hardness and fracturability (crispiness) are very important in judging the magnitude of biscuit quality. Biscuit that exhibits good fracturability trait possess a high degree of hardness and low degree of adhesiveness, which speaks of its crispiness. Crispiness is a good indicator of freshness. Hardness is defined as the maximum peak force during the first compression cycle and has often been substituted by the term firmness. The principal criterion for good biscuit making quality is the diameter increase during the baking process (Rogers *et.al.* 1993). The findings of correlation can be utilized in making a selection programme more effective. The objective of this study was (i) to predict biscuit making quality by using correlation between the most important wheat quality characteristics and (ii) the use of TA.HDi Texture Analyser i.e. a latest technology in determining some other quality attributes of biscuit which can add value to biscuit quality.

MATERIALS & METHODS

The experimental material used in the present study consisted of 50 wheat genotypes, selected from a vast amount of advance breeding material, possessing low protein content (<10.5%) and low sedimentation value (40cc). Samples of these lines were cleaned by air and by hand to remove shriveled and broken kernels and were milled into flour. Other principal tests such as alkaline water retention capacity (AWRC) and the baking test were also applied. Stable Micro Systems TA.HDi Texture Analyser was used to measure the fracturability (crispiness) and hardness of biscuits.

TA.HDi settings were

Mode: Measure force in Compression; **Option:** Return to start; **Pre-test speed:** 1.0 mm/s; **Test speed:** 0.5 mm/s; **Post-test speed:** 10.0 mm/s; **Distance:** 2mm; **Trigger force:** Auto – 5 g; **Data acquisition rate:** 400pps; **Accessory:** 2mm cylinder probe (P/2) using 5 kg load cell; Heavy Duty Platform (HDP/90) with a holed plate.

Sample preparation: Biscuits were taken out of their packets just before testing.

Test set-up: After securing the holed plate into the Heavy Duty Platform, the position of the Heavy Duty Platform was adjusted so that the probe is directly above the hole of the holed plate. Later on calibrations of force and probe were done as per the guidelines of Stable Micro Systems Ltd., U.K. The penetration tests were commenced after the calibrations.

OBSERVATIONS

Once the trigger force of 5g has been attained, the probe proceeds to penetrate the sample to the specified distance. The probe then returns to its start position (Fig.1).

As shown in Figure 2, each biscuit was penetrated at four places and average of these was considered for drawing the conclusions. Curves were plotted with the help of Texture Expert Exceed software. The fluctuations are a result of the biscuit fracturing. The area under the curve is taken as indication of the 'hardness', and the linear distance as an indication of the 'fracturability'.

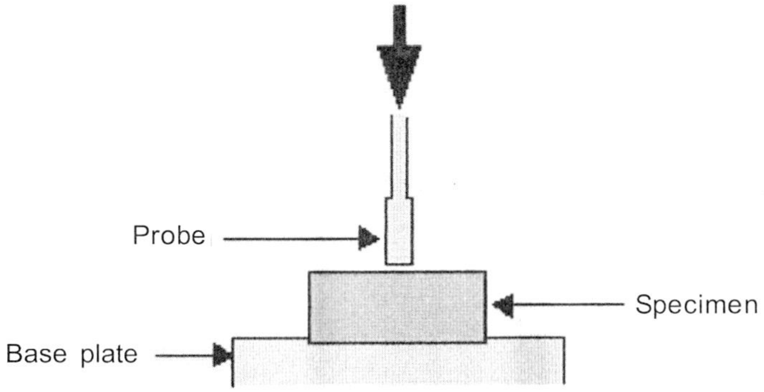

Fig.1: Penetration using a narrow cylindrical probe
(*Source*: Stable Micro Systems)

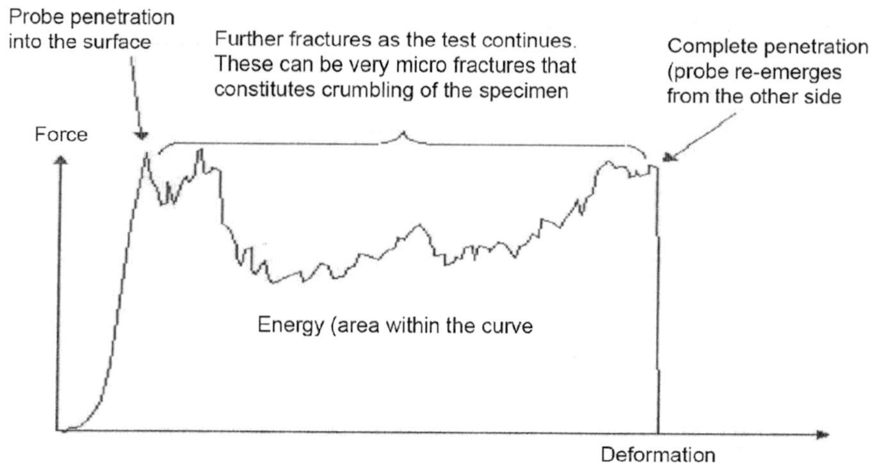

Fig. 2: A typical curve resulting out of probe penetration tests showing hardness and Fracturability

(*Source*: Stable Micro Systems)

EVALUATION & ANALYSIS

Protein content (%)

The protein was estimated using Kjeltec Auto 1030 Analyzer. About 10g seed sample of each genotype was taken and oven dried for 24 hrs. Seeds were ground in fine powder with the help of a grinder. Powdered sample weighing 0.5 g was taken in digestion tubes along with 4.5 g of catalyst mixture (Tablets of $CuSO_4$ + K_2SO_4) and 10 ml conc. H_2SO_4 was added in each tube. The digestion tubes were heated until fumes ceased and simmered briskly. Solution became clear in 15 to 20 min. After cooling 75 ml of distilled water was added and shaken vigorously till the contents became clear. The digested content was distilled with the automatic Kjeltec. Protein value was automatically displayed on the monitor.

The other chemicals required during distillation were as follows:

a. 1% boric acid + mixed indicator (methyl red + bromocresol green)

b. 40% NaOH

c. Distilled water for auto rinsing to avoid any wastage of digested sample.

Sedimentation value (cc)

The sedimentation value was determined following the Sodium Dodecyl Sulphate (SDS) procedure as elaborated by Dick and Quick (1983).

Biscuit spread factor

The estimate of biscuit spread factor provides a basic baking method for evaluating the quality of biscuit flour. The procedure for preparing the sugar-snap biscuit was adopted from Mishra *et al.* (1998). For calculations, biscuits were laid edge to edge and then width was measured. They were rotated one-quarter turn and measured. This was repeated twice. The biscuit width is mean of the four measurements. Finally, biscuit spread factor was derived by dividing diameter of the biscuit with its thickness. Thickness was measured by Vernier Callipers.

Alkaline Water Retention Capacity (%)

The alkaline water retention capacity (AWRC) of flours was determined following the procedure given by Yamazaki (1953).

Fracturability and Hardness

Stable Micro Systems TA.HDi Texture Analyser was used to measure the fracturability (crispiness) and hardness of biscuits. The Linear Distance function calculates the length of an imaginary line joining all points in the selected region. A jagged curve would consequently produce a much larger linear distance value when compared to a smooth curve. The greater the linear distance value the easier the sample is fractured. The Fracturability is a parameter that was initially called as brittleness. It is the force with which a sample crumbles cracks or shatters. The area (g.s) under the curve is taken as indication of the 'hardness', and the linear distance as an indication of the 'fracturability'. Simple correlation was computed among quality traits to determine their interrelationships (Searle, 1961).

RESULTS & DISCUSSION

The application of soft wheat (protein <10%) are very different from those of hard and durum wheat, therefore testing criteria are also significantly different. American Association of Cereal Chemists (AACC) recognizes sugar-snap biscuit spread factor one of the most important baking tests for evaluating soft wheat for their biscuit making qualities. The biscuit quality is considered poor, average, good, very good and excellent when the spread factors are <6.0, 6.1-7.0, 7.1-8.0, 8.1-9.0 and >9.0, respectively (Gupta *et al*, 2002). The wheat genotypes, in the current investigation, fall in these categories were 26, 42, 26, 6 and 0%, respectively. Soft wheat with weak gluten strength and low protein content are the basic quality requirements for making good quality biscuit. The genotypes falling in good to very good category of biscuit spread factor were presented in Table 1 along with other important quality traits.

Low protein content is a good selection criterion for biscuit making quality because low protein is generally combined with weak gluten, which in turn produces more tender and large biscuits. The water insoluble gluten proteins are responsible for the elasticity and cohesiveness (strength) of dough and are important determinants of end use quality products (O'Brien and Ronalds, 1987). Other researchers reported low sedimentation value being good for biscuit making (Knezevic, 1993). Table 1 reveals that all the selected genotypes possess low protein as well as low sedimentation value.

Table 1: Flour protein content, sedimentation value, spread factor, fracturability index and hardness value for selected soft wheat genotypes

Genotype Designation	Protein (%)	Sedimentation Value (cc)	Spread factor (%)	Fracturability index*	Hardness**
98007P - 3B	8.2	40	7.1	22.76	5524
98017P - 6B	8.9	42	7.4	26.04	6971
98072P - 2B	9.0	42	7.3	24.92	7435
98072P - 6B	8.6	41	7.8	21.72	4777
98072P - 7B	9.1	36	7.8	35.02	5744
98072P - 12B	9.2	37	7.9	37.34	2799
98072P - 23B	8.3	40	7.6	36.21	6764
98078P - 2B	8.7	30	8.3	32.66	6870
98088P - 1B	8.9	41	7.7	33.51	8210
98111P - 5B	9.6	40	8.1	25.90	4433
98115P - 4B	8.5	38	7.5	30.57	5813
97002P - 7B	8.7	40	8.1	33.00	7022
97002P - 30B	8.7	42	7.7	29.62	7237
97053P - 12B	8.4	30	7.1	41.83	7922
97053P - 16B	8.6	40	7.2	50.76	12390
97068P - 10B	8.4	33	7.5	32.02	7825

* Correspondents to linear distance in curve; ** Indicates curve area in g.s

It is apparent from Table1 that there is a clear-cut difference among genotypes for fracturability and hardness though they possess low protein, weak gluten and good spread factor. There has been growing concern about the increasingly hard characteristics of genetically soft wheat varieties, as they tend to produce harder products (Wainwright *et al*, 1985). In this context genotype 97053P - 16B was observed to be exceptional as it possesses low protein (8.6%), weak gluten (SDS value 40 cc), very good spread factor (7.2) combined with a very high fracturability index (50.76) and hardness value (12390) (Table 1 & Fig 3).

The genotype 97053P-12B was observed to be superior for fracturability (crispiness) and hardness. Eight other genotypes 97053P-2B, 98072P-23B, 98088P-1B, 98072P-7B, 97002P-7B, 98078P-2B, 97068P-10B and 98115P-4B were found medium in fracturability (crispiness) and hardness. Three genotypes, 98072P-2B, 97002P-30B and 98017P-6B were found good only for hardness. Similarly 98072P-12B was found good only for fracturability (crispiness).

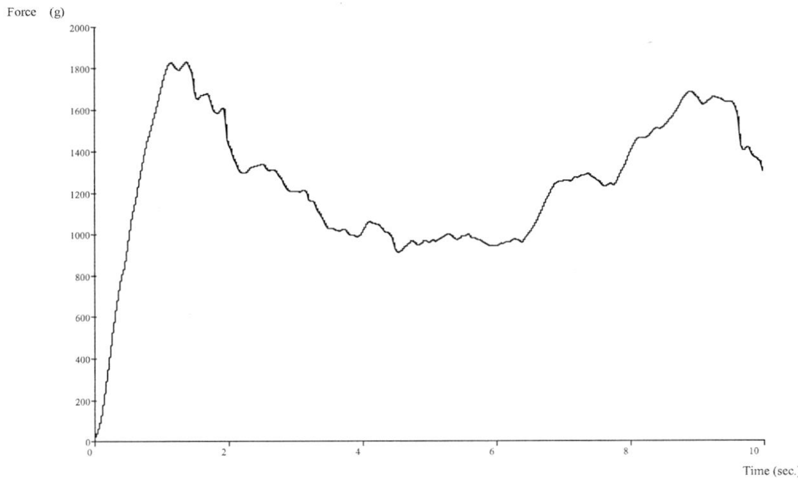

Fig.3: TA.HDi fracture curve of biscuit of wheat genotype 97053P-16B showing hardness and fracturability

Correlation study of biscuit making quality characters

Correlations among biscuit making quality characters viz., protein content, sedimentation value, alkaline water retention capacity, biscuit diameter, thickness, spread factor, fracturability and hardness of 50 soft wheat genotypes were worked out (Table 2).

It reveals from the Table 2 that sedimentation value has very strong correlation with alkaline water retention capacity (r = 0.70). The thickness of biscuit has significant negative correlation with biscuit diameter (r = 0.35). The biscuit spread factor showed negative correlation with protein content, sedimentation value, AWRC, fracturability index and hardness.

Table 2: Correlation coefficient among biscuit quality characters

Correlation Coefficient	Protein (%)	SDS-Value (cc)	AWRC (%)	Dia. (cm)	Thickness (cm)	Spread factor	Fractur-ability index	Hardness (g)
Protein (%)	—	-0.05	0.05	0.16	0.13	-0.02	-0.02	-0.02
SDS-Value (cc)		—	0.70**	-0.22	0.15	-0.17	0.06	-0.20
AWRC (%)			—	-0.21	0.21	-0.24	-0.14	-0.07
Diameter (cm)				—	-0.35*	0.65**	-0.14	-0.01
Thickness (cm)					—	-0.92**	0.18	0.07
Spread factor						—	-0.21	-0.09
Fracturability index							—	0.55**
Hardness								—

* Significant at P=0.05, ** Significant at P=0.01

SDS-value: Sedimentation value, AWRC: alkaline water retention capacity

Biscuit spread factor was taken as an important biscuit quality component. The diameter increase during baking process has been considered as the principal criterion for good biscuit quality. Positive and highly significant correlation was found between biscuit spread factor and biscuit diameter (0.65). This finding has been well supported by O'Brien and Ronalds (1987); Kaldy and Rubenthaler (1987) and Gains (1990). Biscuit spread factor was observed to be negatively correlated with AWRC (r = -0.24), which is supported by the findings of Yamazaki (1953); Labuschagne *et al.* (1996) and Sewaram (2001).

Protein showed non-significant positive correlation of very low order with alkaline water retention capacity (AWRC). Low positive correlation between protein and AWRC is supported by Kaldy and Rubenthaler (1987). However, Gains (1990) suggested that AWRC and protein content are independent variables and have no correlation, which was observed in this investigation also. The fracturability trait showed negative correlation with most of the quality trait. However, it has shown positive correlation with biscuit thickness (r = 0.18) and highly significant positive correlation with hardness.

CONCLUSIONS

The genotype 97053P - 16B was observed to be exceptional as it possesses low protein, weak gluten, and very good spread factor combined with a very high fracturability index and hardness value. The genotype 97053O-12B was observed to be superior for fracturability (crispiness) and hardness in addition to other important quality traits. Eight other genotypes, 97053P-2B, 98072P-23B, 98088P-1B, 98072P-7B, 97002P-7B, 98078P-2B, 97068P-10B and 98115P-4B were found medium both in fracturability (crispiness) and hardness, however they possessed all important biscuit making quality traits. These genotypes could be used as donor for developing good biscuit making varieties in wheat breeding programme. The findings of correlation study may be helpful in selecting good biscuit making genotypes during the selection process.

Like bread, there is a ample scope of improvement in the biscuit quality to bring it to the international level. Measuring hardness and fracturability (crispiness) of biscuits with the help of TA.HDi Texture Analyser adds new dimensions in analyzing the biscuits for their quality. Therefore, it could be concluded that emerging technologies like TA.HDi would help in analyzing wheat genotypes for their suitability for biscuits with added dimensions and it would lead to value addition in the marketability of biscuits.

ACKNOWLEDGEMENTS

We thankfully acknowledge the financial support received from NATP funded project "Quality Improvement of Bread and Durum Wheat – PSR 20" for conducting this investigation.

REFERENCES

Dick, J.W. and Quick, J.S. 1983. A modified screening test for rapid estimation of gluten strength in early-generation Durum wheat breeding lines. Cereal Chemistry, 60:315-318.

Friedman, H. H., Whitney, J. E. and Szczesniak, A.S. (1963). The texturometer - a new instrument for objective texture measurement. J. Food Sci., 28: 390-396

Gains, C.S. 1985. Associations among soft wheat flour particle size, protein, content, chlorine response, kernel hardness, milling quality, white layer cake volume and sugar-snap cookie spread. Cereal Chemistry, 62(4):290-292.

Gains, C.S. 1990. Influence of chemical and physical modification of soft wheat protein on sugar-snap cookie. Dough consistency, cookie size and hardness. Cereal Chemistry, 67(1):73-77.

Gupta, R.K., Sewa Ram and Chauhan, D.S. 2002. Quality of Indian Wheat, Directorate of Wheat Research, Karnal- 132 001, India, Research Bulletin No. 14: pp92.

Kaldy, M.S. and Rubenthaler, G.L. 1987. Milling, baking and physico-chemical properties of selected soft white winter and spring wheat. Cereal Chemistry 64(4): 302-307.

Knezevic, D., Surlan-Momirovic, G. and Ciric, D. 1993. Allelic variation at Glu-l loci in some Yugoslav wheat cultivars. Euphytica. 69:89-94.

Labuschagne, M.T., Brookscoetzee, M.C. and Deventer, S.C.V. 1996. Biscuit-making quality predication using heritability estimates and correlations. J. Sci. Food Agric., 70 : 25-28.

Maache-Rezzoug, Z., Bouvier, J. M., Allaf, K. and Patras, C. (1998). Effect of principal ingredients on rheological behaviour of biscuit dough and on quality of biscuits. Journal of Food Engineering, 35: 23-42

Mishra, B.K., Gupta, R.K. and Sewa Ram. 1998. Protocols for evolution of wheat quality. Directorate of Wheat Research, Karnal. pp 66.

O'Brien and Ronalds, J.A. 1987. Heritabilities of small scale and standard measures of wheat quality for early generation selection. Aust. J. Agric. Res., 38: 801-808.

Preston, K.R., March, P.R. and Tipples, K.H. 1982. An assessment of the SDS sedimentation test for the prediction of Canadian bread wheat quality. Can. J. Plant Sci., 62:545-553.

Rogers, D.E., Hoseney, R.C., Lookhart, G.L., Curran, S.P., Li, W.D.A., Sears, R.G. 1993. Milling and cookie baking quality of near-isogenic lines of wheat differing in kernel hardness. Cereal Chem., 70:183-187.

Searle, S.R. 1961. Phenotypic genotypic and environmental correlations. Biometrics, 17:474-480.

Sewa Ram, Mishra, B.K., Nagarajan, S. and Ram, S. 2001. Biscuit making quality of advanced lines of wheat in India. J. Food Sci. and Tech. 38(5): 476-479.

Wainwright, A.R., Cowley, K.M. and Wade, P. 1985. Biscuit making properties of flours from hard and soft milling single variety wheat. J. Sci. Food Agric. 36: 661-668.

Yamazaki, W.T. 1953. An Alkaline Water Retention Capacity Test for the Evaluation of Cookie baking Potentialities of soft winter wheat flours. Cereal Chemistry, 30: 242-246.

12

Strategies for Nutrient Fortification of Wheat Flour and Bakery Product

Alok Kumar Srivastava

Abstract

The control of micronutrient deficiency is a realizable goal, notwithstanding the magnitude of the task, challenges and constraints that need to be resolved. Since wheat products particularly maida and atta form the major raw material in widely consumed food items like chapati, bread and biscuit. Fortification of flour as followed in many other countries could be an important strategy to help realize this goal. The technology of flour fortification is effective and cost affordable. Effective flour fortification program would require collaboration between scientific community, governmental and non-governmental agencies, and industry and consumer groups.

INTRODUCTION

Malnutrition of larger section of population continues to be the major nutritional problem, which demands attention. In India alone, more than 200 million people in particular women and children suffer from or are at the risk of deficiencies of calorie and various nutrients, particularly vitamins and minerals, commonly referred as micronutrients. Government of India and different State Governments under various welfare programmes have been striving sincerely to improve the nutritional and health status of population.

Apart from major deficiencies of nutrients in the diet, inadequate food supplies, poor health conditions and limited purchasing power, phenomenal increase in population has substantially neutralized the efforts of various nutritional intervention and improvement programmes, resulting in lesser impact. Although most sufferings prevail among poorer section which constitute a sizeable proportion of our population, malnutrition cuts across the rich-poor and rural-urban divide of society. It is estimated that 35-50% of our population does not get one or other dietary nutrients as prescribed in Recommended Daily Allowance. The most frequently encountered deficiencies amongst our population are for Iron, Calcium, Iodine, Vitamin A and B group. Micronutrient malnutrition may lead to grave health, social and economic consequences, which may include learning and neurological disabilities, impaired work capabilities, illness and mortality.

FOOD FORTIFICATION

Global experiences suggest that food based approaches are the most effective in addressing malnutrition and nutrient deficiency disorders. Food based approaches may include dietary diversification, improved food availability, nutrition education and food fortification. Fortification of foods with deficient nutrients is found more effective, economical and holistic long range approach, whereby one or more nutrients can be added to staple food which may not necessitate changes in diet patron of population.

International Status on Food Fortification

Food fortification has played a major role in improving the health and nutritional status in population of developed countries, where there is high dependence on processed foods and industries are streamlined and automatic. Vehicle of food fortification around the globe include oil, dairy products, salts, spices, extruded foods, baby foods, cereal products, tea, etc. Margarine fortified with Vitamin A was introduced in Denmark. Switzerland showed the path for salt iodization for other countries to follow in effectively controlling the iodine deficiencies. Milk was fortified with vitamin A and later flour with iron in number of European countries and North America. In more than 100 countries including developed and developing ones of Latin America, North Africa and Middle Eastern Europe food fortification is commonly practiced today. Many countries have established mandatory standards for fortification of foods with deficient nutrients. Several countries including US have established mandatory standards for fortified flour and bakery products. Guatemala was the first country to include Folic Acid in wheat flour fortification. Food fortification has played major role in improving the health and nutritional status of population of these countries.

Indian Experience in Food Fortification

Food fortification is practiced only to a limited extent in India. The fortification of Vanaspati with Vitamin A (25 I.U./gram) has been a success story. Mandatory enrichment of Vitamin A is also prescribed for Margarine (30 I.U./gram). Iodization of salt (20 ppm potassium iodide) in India has been a success story and it took about a decade to achieve coverage of nearly 80% of total salt consumption. India with production of nearly 4 million tonnes is the second largest producer of iodized salt.

Presently there is no extensive program for fortification of staple foods with micronutrients in India. Considering the wide prevalence of malnutrition in vulnerable population of India, there is urgent need for a sustainable programme for fortifying staple food, in particular, wheat flour with deficient micronutrients.

Wheat Products: Ideal Vehicle for Nutrient Fortification

Various criteria which could be considered during selection of a vehicle for food fortification may include; high coverage of population irrespective of socio-economic status, regular and substantial consumption of staple foods, possibility of centralized production, simple and low cost technology, consumer acceptance of fortified products without effect on quality and storage stability. Cereals are the main source of food for the population of developing countries like India, where over half of calorie intake is derived from cereal grains. Processing of cereals mainly milling reduces level of different micronutrients to varying extent. The nutrient composition of wheat and milled flour (Table 1) clearly indicates 30-70% loss in different nutrients as a consequence of milling of wheat into flour. Fortification not only restores but also improves the nutrient level of flour. Global experiences suggest that among different cereals, wheat flour is the most favoured vehicle for nutrient fortification due to reasons like wide consumption as staple bakery and traditional products by all section of population, ease of fortification and minimal effect on product prepared from fortified flour.

Table 1: Average Nutrient of whole wheat and refined flour

Nutrients	Unit	Whole wheat flour	Refined Flour	% Retention
Water	%	10.3	11.9	-
Energy	Cal	339.0	364.0	105
Protein	%	13.7	10.3	75
Calcium	ppm	340.0	150.0	45
Iron	ppm	38.8	11.0	30
Zinc	ppm	29.3	9.0	30
Riboflavin	ppm	2.2	0.5	30
Niacin	ppm	63.7	18.8	30
Vitamin A	μ gm	64	25	40

Selection of Fortificants

Considerations on the type and level of nutrients for any food fortification should be based on the facts regarding nutritional deficiencies of the population. The most frequently encountered deficiencies amongst our population are for Iron, Calcium, Vitamin A, Folic Acid and B group Vitamins. And fortification of above nutrients to wheat flour supported with appropriately designed interventions hold the potential for alleviating diseases linked to nutritional deficiencies. The technological aspects, which need to be considered for such fortification programmes, are National Product Standards, Fortification process of nutrient and quality control procedures. While choosing a source of fortification, criteria like bio-availability, product compatibility with nutrient source, commercial availability and cost are considered. Further, level of fortification of nutrients is decided considering factors like Recommended Daily Allowance, extent of nutrient deficiency in population, per capita consumption of particular food product, presence and absence of promoter or inhibitor of nutrient absorption.

Iron is the most difficult mineral to fortify to any food. Iron compounds with high bio-availability like ferrous sulfate may cause slight colour or flavour problems in food vehicle, while organoleptically inert compounds are often poorly absorbed. Further food vehicles may contain potent inhibitor of iron absorption, unless manufacturers protect them from inhibitor. Commonly used iron salts and their properties for wheat flour fortification are given in Table 2. Similarly, compounds of other fortifying nutrients which could be considered for wheat flour fortification are given below:

Calcium : Calcium carbonate/chloride/lactate/citrate/phosphate
Zinc : Zinc sulfate
Vitamin A : Vitamin A acetate/palmitate, β carotene
Vitamin B1 : Thiamin chloride/hydrochloride/di phosphate
Vitamin B2 : Riboflavin
Niacin : Niacin Amide

Table 2: Iron salts, and their properties, used in flour fortification

Iron Source	Solubility	Relative Bio-availability	Relative cost
Ferrous sulfate (20%)	Excellent	100	1.0
Ferrous fumarate (33%)	Medium	95	1.2
Reduced iron (97%)	Poor	30-90	0.5 –1.0
Iron EDTA (13%)	Good	200-300	6.0

Nutrition and Food Fortification related Technologies at CFTRI, Mysore

CFTRI with its vision to facilitate providing nutritious foods to population has developed numerous technologies to meet specific nutrition and health requirements of different sections of population in over fifty years of its history. Some of its path breaking research has focused on developing technologies for infant food, weaning food and high-energy multipurpose food. Many Governmental agencies have utilized some of these formulations for mid day meal and school feeding programmes.

Institute was first to recognize bakery and wheat based products as carrier of nutrition in developing products like paushtik atta and high protein 'Nutro' biscuits. CFTRI has developed technology to fortify the wheat flour with vital nutrients at both roller flourmill and small plate mill (chakki) level. Utilizing the existing facilities with minimal additional attachments, wheat flour can be fortified at both levels in simple procedure. Based on the nutrition requirements of vulnerable population and end product compatibility the level and form of different nutrients have been standardized (Table 3).

Table 3: Recommended nutrient addition to wheat flour

Nutrients	Maida (mg/100g)	Atta (mg/100g)	% RDA
Vitamins			
B1	0.65	0.30	40
B2	0.35	0.25	40
Niacin	10.0	5.0	70
Folic acid, μg	50.0	50.0	50
Minerals			
Iron	6.0	8.0	33
Calcium	115	100	33
Zinc	4.0	4.0	33

With the use of fortified flour, the bakery (bread, biscuits, buns and rusk), pasta (Vermicelli) and traditional products (chapati, puri and parotta) can be prepared which have better nutritional value, in addition to their good sensory properties (Table 4). Cost of such nutrient fortification works out to be approx. 30-40 paise/ kg of wheat flour.

Fortification Strategies for Wheat Flour

Fortification of basic food staples needs to be a public health programme. The governmental agencies need to encourage such programmes through legislation and regulation. Further food fortification provides a unique opportunity to industry and trade to play a leading role in health intervention under commercial environment.

There are over 900 roller flour mills in the country, processing over 7 million tonnes of refined flour (made), mainly used for bakery products. In addition, over 50 million tonnes of wheat is processed into whole wheat flour (atta) in plate mills (chakki), numbering over 5 lakh units. Whole wheat flour forms the major raw material for staple foods like chapatti. In recent years greater interest has been shown by the flour millers to market wheat products particularly whole wheat flour in unit packs and this changing trend would help the milling industry to utilize the idle capacity.

Table 4: Quality of bakery and traditional products from fortified flour

	Bakery products						Traditional products				
	Bread				Biscuit		Puri			Chapati	
Fortificant premix level (%)	Specific Volume (ml/g)	Crumb colour "E	Crumb Texture (g force)	Sensory score (max-80)	Breaking force (g force)	Sensory score (7 point)	Oil uptake (%)	Shear Value (g)		Sensory score (max-60)	Value Shear (g)
0	4.32	33.96	240	68.4a	1407	5.85	26.3	2311		55.0a	1055
0.5	4.28	33.03	242	68.4a	1432	5.60	26.8	2380		54.2a	1095
1.0*	4.24	33.37	245	65.4a	1497	5.50	26.5	2405		54.0a	1118
1.5	4.22	32.45	250	62.6b	1548	5.00	26.3	2468		52.2b	1170

*Optimum

Above scenario would also help the flour mills in using wheat products as career of nutrition as end products like bread, biscuit and chapati are consumed by all sections of population. Total cost of such fortification is lower when produced under organized sector on large/mass scale. Technical feasibility of such fortification of wheat products needs to be carefully devised and monitored.

The method of fortification of wheat flour should assure uniform distribution and stability of added nutrients. For flour fortification, dry mixing of nutrients in form of premix into flour stream is commonly practiced. Generally fortification is done at milling stage but nutrients are also fortified to different bakery products at dough mixing stage in batch operations. Soluble sachets of nutrients have also been used at dough stage, which dissolves when water is added. In case of breakfast cereals and snacks, the fortificants are added to extrusion mix.

1. Fortification process at the level of roller flour mill

Since fortification of flour involves more than one nutrient, it is common practice to add all of them in form of single premix. Nutrient fortification to wheat flour basically requires two systems: a feeder to meter-out the fortificant or premix and conveying cum mixing system to ensure uniform blending of nutrients into flour. Among the feeder mechanism generally used, screw feeder is most commonly employed, as it sustains a more constant addition over wider delivery rate, uses fewer mechanical parts and are more sanitary, less expensive and easier to maintain. Revolving disc type feeder has disadvantage of its small size hopper, requiring frequent refill. Micronutrient feeder, having variable feed rate, could be installed over straight run or atta conveyer in roller flour mill. Between the conveying and mixing systems, screw mixers and ribbon blenders are more common, which may work on the gravity feed or pneumatic conveying. Use of entoleter may improve the homogeneity and sterility of fortified milled products.

2. Flour fortification process at the level of plate mill (chakki)

Fortification at plate mill level basically includes simple attachments like micronutrient feeder and paddle conveyor. Delivery rate in variable speed micronutrient feeder could be adjusted to match the flow rate of atta (1:100). Atta along with micronutrients in premix form are simultaneously cooled during its mixing and conveying in paddle conveyor. To further improve the product cooling and nutrient homogeneity, aspiration system and entoleter could optionally be attached to plate mill. Fortification facility at plate mill is simple and compact, which does not require expensive attachments and utilizes latent energy through common shaft to drive micronutrient feeder and conveyor. There is need to fortify folic acid and vitamin A in water soluble, stabilized and dry palmitate form into wheat flour to combat neural tube diseases (NTD) and Vitamin A deficiencies (VAD). Further fortification of Fe EDTA salt to flour could be more bio-available and effective iron source in controlling anaemia, albeit it's higher cost.

Constraints and Challenges

- Lack of awareness about magnitude of nutrient deficiencies and their consequences among various agencies and population, in general.
- Transfer of technology for fortification of wheat flour and bakery/traditional products is essential.
- Emphasis needs to be given in establishing well-defined distribution channel of fortified foods.
- Inhibitor of nutrients like phytates, which reduces the efficacy and bio-availability of fortificants need to be controlled.
- Greater need for periodic monitoring and continual surveillance to ensure safety and sustainability of fortification program of wheat flour and its products.
- Regulatory and legislative support by governmental agencies in order to facilitate the food fortification programmes.
- Awareness about importance of coordination among various concerned groups involving governmental and non-governmental agencies, technologists, and industry and consumer groups.

REFERENCES

Chandrasekhara, M.R. 1984. Fortification of food. In: Interfaces between Agriculture, Nutrition and Food science. Ed. KT Achaya. UNU publications Tokyo, Japan, pp.211-230

Haridas Rao P. and Alok K Srivastava. Micronutrient fortification in roller flour mills. In: Proceedings of National workshop to commemorate Platinum Jubilee of Roller Flour Miller's Federation of India, New Delhi, 9th September 2000.

Haridas Rao, P. and Alok K., Srivastava. 2001. Fortification of cereal and dairy products. In: Need for food fortification in India. Ed. Kalyan Bagchi and Sunder Gujral. Nutrition Syndicate, New Delhi, pp. 49-62.

Prakash, V. Health and better nutrition through adaptable technologies to reach the unreachable. In: Proceedings of National Seminar on Food Fortification in India: Do we need it and why? Nutrition Syndicate. New Delhi, 27-28 April 2000.

Prakash, V. Role of food processing and preservation in preventing micronutrient malnutrition. In: Proceedings of workshop on prevention and control of micronutrient malnutrition through food based actions in SAARC countries. Dhaka, FAO Publication. pp. 67-70.

Ranum, P. 2000. Fortification of high-extraction wheat flours. Cereal Foods World, 45(6): 267-268

Ranum, P. A 1998. Dialog on enriched white flour and bread. Cereal Foods World, 43(5): 384

Ranum, P. Cereal enrichment. In Lorenz, K. J., Kulp, K. Ed. Handbook of Cereal Science and Technology. Dekker, New York, NY, USA. 1991

Venkatesh Mannar, M.G and Indira Chakravarty. Fortification of flour to control micronutrient deficiencies. In: Proceedings of National Seminar on wheat and wheat products – Development Strategies for the Decade. Chennai. 1998, April 17-18. pp. 169-173.

13

Value Addition in Wheat Through Identification of Genotypes for Biscuit Making Qualities

Archana Vaishnava, Sohan Pal, D.P. Saini and J.P.Jaiswal

Abstract

30 lines were selected from 864 germplasm lines for good biscuit making attributes on the basis of low protein content (<10.5%) and low sedimentation value (<40cc). These lines were further screened by actual baking test and alkaline water retention capacity (AWRC) test keeping the standard prescribed for biscuit making qualities viz., low protein content, low sedimentation value, low AWRC and larger diameter with larger spread ratio of biscuit. AWRC had significant negative correlation with biscuit spread ratio. Positive significant correlation was found between biscuit spread ratio and biscuit diameter. On the basis of above findings, it may be recommended that low protein content, low sedimentation value, low AWRC may be used as criteria for better biscuit making i.e. increase biscuit spread ratio and biscuit diameter. So wheat-breeding programme has opened up new opportunities for the end use quality product and wheat based food industry. Hence there is an urgent need for qualitative analysis of desired traits in wheat towards enhanced value addition and thereby an attempted has been made to identify suitable genotypes possessing good biscuit making attributes.

INTRODUCTION

Wheat falls in a distinct market class and have milling and baking properties conditioning them for specific products (Yamazaki et.al 1981). The global demand for wheat is increasing as a consequence of rapid population growth and changing consumer taste. In view of this, there is need to produce good quality wheat for good chapattis and other bakery products. In this study only biscuit-making quality is the diameter increase during the baking process. A large diameter is considered superior (Rogers et.al. 1993). All characteristics associated with biscuit diameter, as measured by the sugar-snap biscuit test were given due consideration in this study. Correlation, although reliable only for the range of material tested, may point to relationships that can be utilized in making a selection programme more effective. A correlation coefficient between two quality attributes that is usually high suggests a strong heritable association and possibility a narrow genetic base. Very little information is available to plant breeder on correlation of wheat quality traits, especially for biscuit making quality traits. The aim of this study was to predict biscuit making quality by using correlation between the most important wheat quality characteristics.

MATERIALS AND METHODS

The experimental material used in the present study consisted of 30 lines of wheat possessing low protein content (<10.5%) and low sedimentation value (40cc.). Samples of these lines were cleaned by air and by hand to remove shriveled and broken kernels and were milled in to flour. Other principal tests such as alkaline water retention capacity (AWRC) and baking test were also applied for these 30 lines. The averages of biscuit concerning character are given in Table 1.

Evaluation and Analyses

Wheat lines were evaluated for protein content and sedimentation value. Other principal tests such as alkaline water retention capacity (AWRC) and increase in diameter and biscuit spread ratio during baking were also applied. A simple correlation coefficient was computed among all characteristics.

Protein content (%)

The protein was estimated using Kjeltec Auto 1030 Analyzer. About 10 g seed sample of each genotype were taken and oven dried for 24 hrs. Seeds were ground in fine powder with the help of a grinder. Powdered sample weighing 0.5 g was taken in digestion tubes along with 4.5 g of catalyst mixture (Tablets of $CuSO_4 + K_2SO_4$) and 10 ml conc. H_2SO_4 was added in each tube. Digestion tubes were heated until fumes ceased and simmered briskly. Solution became clear in 15 to 20 min after cooling 75 ml of distilled water was added and shaken toughly till the contents became clear. The digested content was distilled with the automatic Kjeltec. Protein value was automatically displayed on the monitor.

Other chemicals required during distillation were as follows:

a. 1% boric acid + mixed indicator (methyl red + bromocresol green)

b. 40% NaOH

c. Distilled water for auto rinsing to avoid any wastage of digested sample.

Sedimentation value (cc)

The sedimentation value was determined following Sodium Dodecyl Sulphate (SDS) procedure (Dick and Quick, 1983).

Biscuit spread ratio

The estimate of biscuit spread ratio provides a basic baking method for evaluating the quality of biscuit flour. The procedure involved as suggested by Mishra et al. (1998) for preparing two sugar snap biscuit is given below:

Sugar (24.0 g), nonfat dry milk (1.20 g) and sodium bicarbonate (0.40 g) were sifted together and creamed with shortening (12.0 g). 37.6 g of this creamed mass was weighed and put into a mixing bowl. 4.0 ml of solution A (79.8 g sodium bicarbonate dissolved in distilled water, final volume 1 litre), 2.0 ml solution B (101.6 g ammonium chloride and 88.8 g sodium chloride dissolved in distilled water, final volume 1 litre) and appropriate amount of water were added to achieve optimum dough consistency. 40 g flour was added to the mixing bowl and mixed. The dough was scraped from bowl and cut into two equal portions and these were transferred to lightly greased sheet. Using gauge strips, these were rolled to thickness with one forward rolling pin stroke and then return (backward) stroke. The dough thus obtained was cut with cutter, discarding excess dough. It was immediately placed in oven (205°C, 11 minutes). After removal from oven the biscuits were allowed to cool down and after that biscuit spread was measured.

For calculations, biscuits were laid edge to edge and width was measured. They were rotated one-quarter turn and measured. This was repeated twice. The biscuit width is mean of the four measurements. Finally biscuit spread ratio was obtained by dividing the biscuit diameter with its thickness. Thickness was measured by Vernier Callipers.

Alkaline water retention capacity (%)

The alkaline water retention capacity (AWRC) of flours was determined following the procedure given by Yamazaki (1953). Fifteen-gram flour (14% moisture basis) was weighed into a weighed 100 ml centrifuge tube with stopper. Fifty ml 0.1 N sodium bicarbonate solution was added to this. It was closed with stopper and shaken vigorously to suspend flour. The stopper was removed and 25 ml more of the same solution was added. The solutions were mixed and time noted. Thereafter the tube was shaken vigorously and allowed to stand for 20 minutes, shaking every 5 minutes. The stopper was removed and the content was centrifuged for 15 minutes at 1800 rpm. The supernatant liquid was decanted and the tube was drained for 10 minutes at an angle of 10-15°. The tube with stopper was weighed with contents and the gain in weight was expressed in percent. Care has to be taken on centrifuge time because longer centrifuge time appeared to distort results since samples had different packing or compression rates, and shorter times resulted in incomplete sedimentation in some cases. In our experiment, the contents used were reduced to 1/15 of the original amount. Simple correlation coefficient was computed among quality traits to determine their interrelationships (Searle, 1961).

RESULTS AND DISCUSSION

The main aim of this investigation was to select some lines, good for biscuit purpose, which could be used as donors in a crossing programme for quality improvement. These lines for biscuit making quality were used for further observations such as actual baking test and alkaline water retention capacity test following the biscuit making quality on the basis of low protein content, low sedimentation value, a low AWRC and a larger diameter and spread ratio of biscuits (Rogers *et al.*, 1993).

Few genotypes among the 30 selected germplasm lines, with protein content close to 10% showed maximum spread ratio and larger diameter of biscuits and could be considered suitable for biscuit quality. Low protein content is a good selection criterion for biscuit making quality as low protein is generally combined with weak gluten, which in turn produces more tender and large biscuits than those produced from higher content of protein and strong gluten.

The water insoluble gluten proteins are responsible for the elasticity and cohesiveness (strength) of dough and are important determinants of end use quality products (O'Brien and Ronalds, 1987). From the total germplasm screened, 30 lines were selected with low sedimentation value (below 40 cc). UPAGP 6, UPAGP 15, UPAGP 301, UPAGP 362, UPAGP 2378, UPAGP 2484, UPAGP 2532, UPAGP 2667, UPAGP 2673, UPAGP 2677 and UPAGP 2654 showed correspondingly increased diameter and high spread ratio of biscuit (Table 1). Low sedimentation value being good for biscuit making has also been supported by other researchers (Knezevic, 1993). Some selected lines showed inferior results for biscuit quality on the basis of low sedimentation value. This may be attributed to flour particle size and the method of determining the sedimentation value (Gains, 1985).

UPAGP 6, UPAGP 15, UPAGP 265, UPAGP 2523, UPAGP 2532, UPAGP 2667, UPAGP 2673, UPAGP 2677 and UPAGP 2654 showed low alkaline water retention capacity, increased diameter and higher spread ratio of biscuit (Table 1).

Table 1: Values of different biscuit quality characters in 30 selected germplasm lines

S. No.	Genotypes content (%)	Protein Vol. (cc)	Sed. (%)	AWRC ratio	Spread (cm)	Dia. (cm.)	Thickness
1.	UPAGP 6	10.36	38	69.73	5.7	6.85	1.20
2.	UPAGP 15	10.28	38	77.28	7.3	7.67	1.05
3.	UPAGP 24	10.35	33	84.73	5.2	6.50	1.25
4.	UPAGP 97	9.87	39	78.06	4.5	6.25	1.40
5.	UPAGP 145	9.56	38	82.47	5.4	6.70	1.25
6.	UPAGP 162	9.82	33	79.53	5.4	6.50	1.20
7.	UPAGP 182	9.66	31	73.26	6.1	6.75	1.10
8.	Raj 3765	8.15	38	72.67	5.1	5.90	1.15
9.	UPAGP 265	10.10	38	75.71	6.2	6.85	1.10
10.	UPAGP 301	10.22	32	86.39	5.1	5.90	1.15
11.	UPAGP 362	10.32	38	88.84	5.0	6.55	1.30
12.	UPAGP 2378	10.50	39	96.78	5.89	6.75	1.15
13.	UPAGP 2484	10.25	40	87.75	5.2	6.85	1.25
14.	PDW 215	9.84	34	86.39	5.1	6.75	13.0
15.	UPDGP 716	9.94	29	104.33	5.8	6.75	1.15
16.	UPAGP 2519	10.03	34	81.98	5.5	6.60	1.20
17.	UPAGP 2523	10.44	35	76.20	5.2	6.60	1.27
18.	UPAGP 2524	10.47	35	82.86	5.4	6.45	1.20
19.	UPAGP 2532	10.32	37	75.81	5.7	6.55	1.15
20.	UPDGP 609	10.32	31	82.47	4.7	6.15	1.30
21.	UPDGP 616	9.91	27	95.50	5.1	6.85	1.35
22.	UPDGP 621	10.38	21	93.82	4.9	6.90	1.40
23.	UPDGP 627	10.00	25	83.47	4.8	6.30	1.30
24.	UPDGP 658	9.72	25	79.73	5.6	6.20	1.11
25.	UPAGP 2667	9.69	25	62.77	6.0	6.50	1.08
26.	UPAGP 2673	10.35	38	68.26	6.3	6.50	1.04
27.	UPAGP 2677	9.94	35	70.12	5.5	6.15	1.12
28.	UPAGP 2680	9.34	35	70.12	4.8	6.50	1.35
29.	UPAGP 2654	10.47	40	84.63	5.6	6.35	1.13
30.	UPAGP 2662	10.19	38	114.10	5.0	6.35	1.27

The low AWRC possessed less water absorption which favoured weak dough strength. Weak dough strength is good for biscuit quality. Thus, low protein, low sedimentation value and low AWRC are desirable for high biscuit spread ratio, larger biscuit diameter and less thickness.

Biscuit spread ratio was significantly negatively correlated with AWRC (r=-0.365). This fact has been documented earlier by Yamazaki (1953); Labuschagne *et al.* (1996); Sewa Ram (2001). The AWRC test is actually a standardized method of measuring the water retention ability of flour against centrifugal force. It is well established that lower the percent of water retained better will be the quality (Kaldy and Rubenthaler, 1987). The diameter increase during baking process has been considered as the principal criterion for good biscuit quality. Positive significant correlation was found between biscuit spread ratio and biscuit diameter (r=0.48). This finding is very well supported by O'Brien and Ronalds (1987); Kaldy and Rubenthaler (1987), Gains (1990) and Barveau *et al.* (2001).

Protein showed non-significant positive correlation of very low order with alkaline water retention capacity (AWRC %) and also with spread ratio. Low positive correlation between protein and AWRC is supported by Kaldy and Rubenthaler (1987). However, Gains (1990) suggested that AWRC and protein content are independent variables and have no correlation. As shown in Table 2, on the basis of simple correlation, the sedimentation value was negatively correlated with AWRC (r = -0.126) and weakly positively correlated with spread ratio of biscuit (r = 0.127). The inability of sedimentation value to be correlated with AWRC and biscuit spread ratio should be viewed in the light of the fact that the correlation coefficient between SDS-sedimentation values and the quality parameter are highly dependent on environment (Preston *et al.*, 1982).

Table 2: Correlation coefficient of quality components among themselves

Characteristics	Protein (%)	Sedimentation value (cc)	AWRC (%)	Spread ratio	Biscuit diameter (cm)
Protein (%)	–	0.320	0.204	0.04	-0.023
Sed. value (cc)		–	-0.126	0.127	0.034
AWRC (%)			–	-0.365*	0.085
Spread ratio				–	0.481**
Diameter (cm)					–

* Significant at P= 0.05, ** Significant at P= 0.01

CONCLUSIONS

UPAGP 6, UPAGP 15, UPAGP 301, UPAGP 265, UPAGP 2532, UPAGP 2667 and UPAGP 2673 with low protein, low sedimentation value low AWRC and greater spread ratio lines may be suitable for better biscuit quality as compared to the other selected lines. On the basis of correlation among biscuit quality components, protein content (%), sedimentation value (cc), AWRC (%), biscuit spread ratio and biscuit diameter (cm) and their inter-relationship, three characters i.e. were found of relevance for biscuit quality AWRC, biscuit spread ratio and biscuit diameter. On the basis of above findings, it may be recommended that low protein content, low sedimentation value, low AWRC may be used as criteria for better biscuit making i.e. increase biscuit spread ratio & biscuit diameter. So, Wheat breeding programme has opened up new opportunity for the end use quality product.

REFERENCES

Dick, J.W. and Quick, J.S. 1983. A modified screening test for rapid estimation of gluten strength in early-generation Durum wheat breeding lines. Cereal Chemistry, 60:315-318.

Gains, C.S. 1985. Associations among soft wheat flour particle size, protein, content, chlorine response, kernel hardness, milling quality, white layer cake volume and sugar-snap cookie spread. Cereal Chemistry, 62(4) : 290-292.

Gains, C.S. 1990. Influence of chemical and physical modification of soft wheat protein on sugar-snap cookie. Dough consistency, cookie size and hardness. Cereal Chemistry, 67(1): 73-77.

Kaldy, M.S. and Rubenthaler, G.L. 1987. Milling, baking and physical-chemical properties of selected soft white winter and spring wheats. Cereal Chemistry, 64(4): 302-307.

Knezevic, D., Surlan-Momirovic, G. and Ciric, D. 1993. Allelic variation at Glu-l loci in some Yugoslav wheat cultivars. Euphytica. 69: 89-94.

Labuschagne, M.T., Brookscoetzee, M.C. and Deventer, S.C.V. 1996. Biscuit-making quality predication using heritability estimates and correlations. J. Sci. Food Agric., 70 : 25-28.

Mishra, B.K., Gupta, R.K. and Sewa Ram. 1998. Protocols for evolution of wheat quality Directorate of Wheat Research, Karnal. pp 66.

O'Brien and Ronalds, J.A. 1987. Heritabilities of small scale and standard measures of wheat quality for early generation selection. Aust. J. Agric. Res., 38: 801-808.

Preston, K.R., March, P.R. and Tipples, K.H. 1982. An assessment of the SDS sedimentation test for the prediction of Canadian bread wheat quality. Can. J. Plant Sci., 62:545-553.

Rogers, D.E. Hoseney, R.C., Lookhart, G.L., Curran, S.P., Li, W.D.A, Sears, R.G. 1993. Milling and cookie baking quality of near-isogenic lines of wheat differing in kernel hardness. Cereal Chem., 70:183-187.

Searle, S.R. 1961. Phenotypic genotypic and environmental correlations. Biometrics, 17 : 474-480.

Sewa Ram, Mishra, B.K., Nagarajan, S. and Ram, S. 2001. Biscuit making quality of advanced lines of wheat in India. J. Food Sci. and Tech. 38(5): 476-479.

Yamazaki, W.T. 1953. An Alkaline Water Retention Capacity Test for the Evaluation of Cookie baking Potentialities of soft winter wheat flours. Cereal Chemistry, 30: 242-246.

Yamazaki, W.T., forele, K.W., Kingswood and Greenwood, C.T. 1981. Soft wheat production. In: W.T. Vamazaki and C.T. Greenwood (eds.), soft wheat production, breeding, milling and uses. pp. 1-32. Am. Assoc. Cereal Chem. St. Paul. MN.

14

Effect of Specific Gravity Separation on the Quality of Paddy (*Oryza sativa* L.) Seed During Storage

Satish Kumar, Sadhna Arora, V. K. Sehgal and Sapanpreet Kaur

Abstract

Paddy seeds of variety PR - 116 were obtained from the PAU seed farm. These seeds were cleaned with commonly used seed processing equipment. Samples were divided into 2 parts. One part was kept as such for taking the observations of feed (control) and the remaining part was passed through specific gravity separator for its refinement. The various fractions obtained were recovery, middlings, and rejections. These graded fractions as well as the feed were packed in 5-kg cloth bags and stored under ambient conditions. Different quality parameters viz. moisture content, physical purity, thousand grain weight, percent germination, root/shoot length, vigour index and speed of germination were recorded at 30 days interval. The physical purity, thousand grain weight, percent germination, root length, shoot length and vigour index were found to be higher in recovery and minimum for rejections while moisture content was highest for rejection as compared to all other fractions. Germination, root/ shoot length, vigour index and speed of germination increased with the duration of storage while moisture content, thousand-grain weight and physical purity followed a reversed trend. Compared to feed, the recovery had higher quality attributes after 4 months of storage viz. moisture content decreased by 0.3%; physical purity increased by 0.67%; thousand grain weight increased by 2.97 gm/ thousand grain, germination increased by 8%; vigour index increased by 481.1 points; speed of germination increased by 3.2 seedlings per day. The grading resulted in improvement in overall quality of sunflower seed.

INTRODUCTION

Paddy (Oryza sativa. L) is an important crop widely grown in India. Its production has witnessed a quantum increase in the recent past. Paddy is grown in an area of about 43 million hectares with irrigation coverage of only 45 percent. The area under paddy in Punjab was 2.49 million hectares with total production of about 8.82 million tonnes in 2001-2002 (Anon., 2002). Seed is the most important single basic input, which influences the better utilization of other inputs like fertilizers, insecticides, irrigation water etc. A poor quality seed results into an unsatisfactory stand establishment and poor yields in the field. A seed should have good germination potential and vigour in order to have better stands in the field. Also seeds should be properly cleaned and graded before keeping them for storage. Although wide variety of cleaners-cum-graders are available but the cleaning and grading of seed is commonly done by double screen cleaner-cum-grader coupled with indented cylinder separator. These are unable to separate infested grains or heavy imputations like stone, which look like grains i.e., having identical shape and size. Specific gravity separator suits to these requirements as it separates on the basis of unit weight (density), which is higher for sound grains compared to infested grains.

Flower and Lim (1959) observed that the effectiveness of separating increased upto maximum of about 15° slope and there was a decrease in effectiveness with higher angles. The effectiveness of separation increased with decrease in feed rate in screen cleaners. Patil and Sarode (1988) found that heavy and medium weight fractions of wheat seed from specific gravity separator were superior to the light fractions in percent germination and seed vigour index. Meharwade *et al* 1992 observed that the average seed germination and quality was best in healthy, heavier (graded) seeds and worst in lightweight, undersized seeds. Prasad *et al* (1994) observed that sunflower seeds with specific gravity greater than 0.6 were superior in most quality attribute viz. seed vigour, percent germination, germination rate, and electrical conductivity. Uppal (2002) using specific gravity separator, graded sunflower (PSFH-67) seeds into different fractions and studied the effect of storage on quality attributes in terms of physical purity, moisture content, thousand grain weight, percent germination and vigour index. In view of above facts, the present study was undertaken to meet the following objectives:

i. To study the effectiveness of specific gravity separator with regard to refinement of seed quality.

ii. To study the seed quality parameter viz. physical purity, moisture content, thousand-grain weight, percent germination, vigour index and speed of germination during storage.

MATERIALS AND METHODS

Paddy of variety PR - 116 seeds were obtained from the PAU seed farm. These seeds were cleaned with usual seed processing equipment. Samples were divided into 2 parts. One part was kept as such for taking the observations of feed (control) and the remaining part was passed through specific gravity separator for its refinement. The various fractions obtained were recovery (heavier, healthy seeds), middlings (intermediate, under size) and rejections (lightweight seeds, weevil eaten, and foreign matter). The respective percentages of each fraction were computed. These graded fractions as well as the feed were packed in 5-kg cloth bags and stored under ambient conditions. Different quality parameters viz. moisture content, physical purity, percent germination, thousand grain weight, vigour index and speed of germination were recorded for all the treatments at 30 days interval from Dec., 2002 to April, 2003. Two replications from each treatment were taken. Moisture content was determined by standard oven method (Chakravarty, 1995). Physical purity in percent was calculated by taking 120 gm samples and separating the inert matter, damaged seeds and pure seeds from the samples. Thousand seeds were counted on electronic balance separator and the average thousand grain weight was computed in terms of gm/ 1000 grains. For calculating percent germination, 100 seeds were taken and placed in between moist germination papers with sufficient space between them. These were kept at a temperature of 25°C in an incubator. Samples were taken out after 14 days interval and the no. of germinated seeds were counted and germination percent was computed. From the above-mentioned sample, 15 germinated seeds were taken and their average root length and shoot length was measured in centimetres. Vigour index was calculated using the formula.

Vigour index = [Average root length (cm) + Average shoot length (cm)] X Germination (%)

For finding out speed of germination, four samples of 25 seeds each were taken and placed in petri dishes laid with moist germination paper. No. of seedlings emerging daily were counted from the day of placing the seeds till germination became constant. Speed of germination was calculated as follows (Aggarwal, 1980):

$$\text{Speed of germination} = \Sigma n/d$$

Where n = No. of seedlings emerging daily

d = Days after placing the seeds for germination till it becomes constant

Description of specific gravity separator

Specific gravity separator consists of a triangular shaped perforated table having reciprocating motion. The table is tilted towards the front and towards the left such that a sphere placed on it would roll in the direction of gravity flow. The material is introduced from feed box. The air is blown through the perforated floor at such a rate that the material is partially lifted from the contact with the floor. The lighter & smaller particles are lifted somewhat higher and float down the table towards the discharge end. The heavier and longer particles are not lifted by the air. The oscillating motion moves them in the direction of conveyance and is discharged at the right end of the table. Other materials that are partially lifted are discharged at intermediate points. The steep table pitch can permit high capacity. It is especially useful for removing lightweight infertile, infested seed from seed stock, which cannot be separated by other devices

RESULTS AND DISCUSSION

Machine Parameters

The deck area was found to be 0.1368 m^2. The airflow rate was fixed at 395m^3/hr. The end to end slope and side to side slope were 5:1 and 11:1 respectively. Paddy seeds were fed at the rate of 40 kg/hr. All these machine parameters were chosen to have the best performance from specific gravity separator in separating the seed fractions.

Grade Fractions

Samples of paddy (Variety: PR-116) were divided into two parts; 40 kg was fed to the specific gravity separator for its refinement and the remaining 10 kg was stored as such (control). The different fractions obtained from specific gravity separator were found to be approximately 70, 23, and 7 percent for recovery, middlings, and rejections respectively. Separation of paddy seeds into various fractions was quite effective. However, lowering the middlings percentage is suggested to obtain better refinement.

Moisture Content (%)

It was found that moisture content for rejections was highest in the beginning (11.9 percent) and in remaining treatments it was more or less same (11 percent). A decrease of 2.4 percent moisture content during storage period was observed in all treatments except in rejections where decrease was 1.8 percent. The decrease in moisture content could be attributed to the drier ambient conditions prevailing during storage period. The lower decrease in moisture content in rejections is attributed to presence of stones and other inert matter. This is also verified by a decrease in 1000-grain weight.

Physical Purity (%)

Physical purity was more or less constant during entire period of storage in paddy seeds. The physical purity was highest in recovery followed by feed and middlings and least in rejections. These results also supplemented the effectiveness of specific gravity separator in separating out lightweight, weevil eaten seeds from heavy and healthy seeds. The physical purity stagnated around 99 percent for recovery and around 98 percent for feed and middlings, whereas rejections showed less than 90 percent physical purity during four months storage.

Thousand Grain Weight (gm)

Thousand grain weight decreased marginally with the period of storage in all treatments. The decrease in the weight is attributed to the drier environmental conditions during the entire storage period resulting in a loss of moisture to the environment. The 1000-grain weight was found highest in recovery and lowest in rejections in both seeds, as the rejections contained lightweight or weevil eaten seeds and brokens.

Percent Germination (%)

It was observed that the germination percentage was maximum for recovery and minimum for rejection among all the treatments during entire period of storage (Fig.1). The important observation was that germination potential increased with the storage. The percent germination was highest in the recovery as it contained more healthy, heavy and sound seeds, which was the result of proper grading whereas rejections contained immature and unhealthy seeds. The germination percentage in the paddy seeds was low initially, as the seeds were dormant and then with the storage it increased as the dormancy was overcome (Kapur *et al*, 1988). Among the graded fractions of paddy an increase of nearly 11.5 percent was observed in feed, recovery, and 15 percent in middlings over the extended storage period.

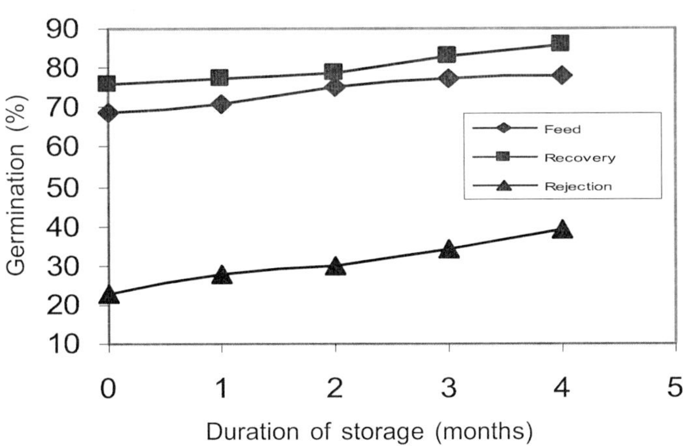

Fig.1: Percent germination in different seed fractions of paddy

Shoot Length and Root length

It was observed that shoot length and root length was maximum for recovery and minimum for rejections. The feed had intermediate values at all periods of storage because it had a mixed lot. It was further observed that shoot length and root length increased with time as the germination percentage increased. The feed accounted for 29.6 percent / 28.1 percent increase in the shoot length/ root length and recovery had 34.16 percent / 45.16 percent increase in the shoot length/ root length.

Vigour Index

The vigour index was highest in recovery and least in rejections among all the treatments throughout the storage period. This is attributed to the fact that vigour index depends upon the percent germination, shoot length and root length, which were higher in recovery compared to rejections (Fig.2). In case of paddy, vigour index increased with the duration of storage as the percent germination increased. There was 10.22 increase in vigour index in feed and 9.3 percent increase in recovery during the first month of storage. Middlings showed intermediate increase (8.4%) in the first month of storage. The vigour index was found to have poor values in case of rejections as it contained light seeds, inert matter, brokens, weevil eaten seeds etc.

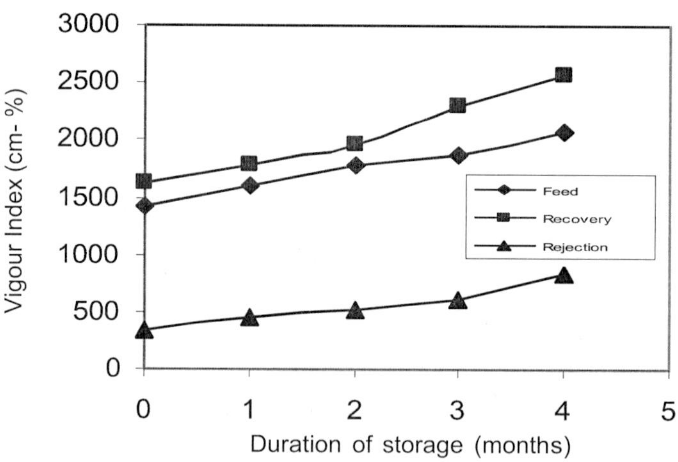

Fig. 2: Vigour Index on different seed fractions of paddy

Speed of Germination

As seen from the Fig.3, the recovery exhibited best results for the speed of germination in paddy seeds Due to the increase in germination potential of the paddy seeds, speed of germination also increased with the duration of storage.

The speed of germination in recovery had maximum increase of 18.6 percent followed by feed and middlings, which had nearly same 15 percent, increase. But in case of rejections slightly more increase (16.1 percent) was observed.

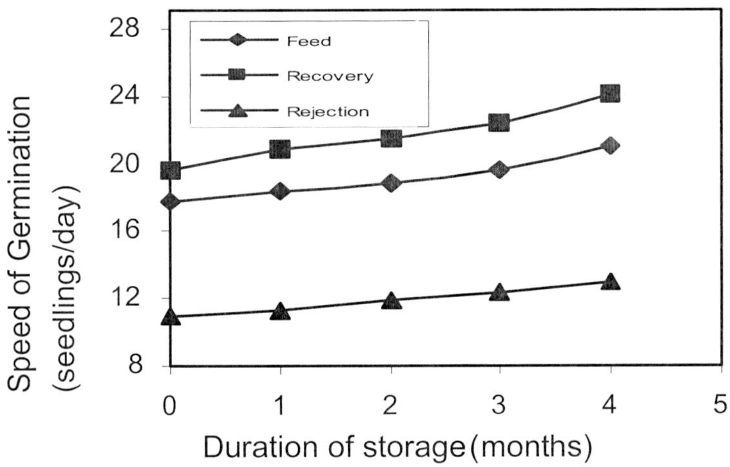

Fig. 3: Speed of germination in different seed fractions of paddy

The following conclusions were drawn from the study (Table 1).

Table 1: Summary of results

Quality attribute	Feed	Recovery	Comparative improvement
Moisture content (%)	9.0	8.3	0.7
Physical purity (%)	98.21	98.58	0.67
Thousand grain weight	24.67	27.64	2.97
Germination (%)	78.0	86.0	8.0
Root length (cm)	12.82	14.69	1.87
Shoot length (cm)	13.88	15.12	1.24
Vigour index (points)	2082.6	2563.7	481.1
Speed of germination (seedlings/ day)	21.0	24.2	3.2

The physical purity, thousand grain weight, percent germination, root length, shoot length and vigour index were found to be higher in recovery and minimum for rejections while moisture content was highest for rejection as compared to all other fractions. Germination, root/ shoot length, vigour index and speed of germination increased with the duration of storage while moisture content, thousand-grain weight and physical purity followed a reversed trend. Compared to feed, the recovery had higher quality attributes after 4 months of storage viz.

moisture content decreased by 0.3%; physical purity increased by 0.67%; thousand grain weight increased by 2.97 gms/ thousand grain, germination increased by 8%; vigour index increased by 481.1 points; speed of germination increased by 3.2 seedlings per day. The grading resulted in improvement in overall quality of sunflower seed.

REFERENCES

Aggarwal RL (1980) Seed technology. Oxford & IBH Publishing Co. Pvt. Ltd., New Delhi.

Anonymous (2002) Package of practices, Punjab Agricultural University, Ludhiana

Chakraverty, A (1995). Postharvest technology of cereals, pulses and oilseeds (3rd edn.). pp: 26-27, Oxford & IBH Publishing Co. Pvt. Ltd. New Delhi

Flower BT and Lim SC (1959). The influence of various factors on effectiveness of separation of finely divided solids by a vibrating screen. Chemical Engg. Sciences, 10:163-170.

Kapur A., Singh H., and Sharma, HL. (1988). Roll of colour gumella in partial dormancy of fodder sorghum seeds. Current Sci. 57(16):888.

Meharwade MM, Giriraj K, Shahidhara SD (1992) Influence of Seed Size, Seed quality and germination in sunflower. Karnataka Journal of Agricultural Sciences, 5(1):73-76.

Patil VN and Sarode SD (1988) Seed Quality Studies in Wheat as Influenced by Specific Gravity Separator. Seed Research 16:1:114-116.

Prasad SR, Reddy A S and Seenappa K (1994) Influence of Specific Gravity on Vigour and Viability of Sunflower seeds. J Oil Seeds Research 11:2, 196-200.

Uppal, Sanjeev (2002) Improvement of sunflower seed quality in marginal lots by specific gravity separator. B. Tech Project Report, Punjab Agricultural University, Ludhiana.

15

Effect of Pea Grits on the Extrusion Cooking of Rice-Pea Blends

Baljit Singh and H.P.S. Nagi

Abstract

Garden peas of two cultivars (Pb-87 and Pb-88) were procured. Pods were shelled manually, grains were blanched and dehydrated. Other lot of peas of both the varieties was allowed to mature on the plant till the plant foliage dried up, harvested, pods were dried in shade for few days, shelled manually and grains were further dried in shade. Grits were prepared by grinding rice and pea (dhal) in a laboratory grinder to pass through 20 mesh sieve. Blends were prepared by mixing rice and pea grits in the ratio of 100:0, 85:15 and 70:30. The blends were subjected to extrusion cooking using Brabender single screw extruder. The extruder die pressure, specific energy consumption (SEC), expansion ratio, density, water absorption index (WAI), water solubility index (WSI) and organoleptic characteristics were studied. Addition of pea grits resulted in a decrease in die pressure in all the blends. Addition of pea grits showed a significant effect on SEC. A decrease in expansion with increase in the level of pea grits in blend was observed. The extrudates having lower expansion showed higher density values and vice versa. Effect of replacement of rice grits with pea grits on WAI was found to be significant. Level of pea grits significantly affected the WSI. Incorporation of pea grits in the rice grits significantly affected the sensory characteristics. Replacement of rice grits with mechanically dried pea grits resulted in deterioration of acceptability score, but addition of field dried pea grits improved the same.

INTRODUCTION

Fresh peas contain high moisture (75-80%) and are thus extremely perishable. The peas remain available to the consumer for a short period (2-3 months). So the processing of peas is very important to enhance its shelf life and availability to the consumer. Although freezing, dehydration and canning of peas is gaining popularity, but even then seasonal glut of peas causes uneconomical return to growers. Therefore, there is a need to develop non-conventional products.

Pea is an important source of nutrients particularly proteins and essential amino acids. Blending peas with cereals can complement each other so that protein in the resulting product more nearly approximates that of a complete or balanced protein. Pea also contains anti nutritional factors which can be inactivated to a large degree by appropriate heating and processing during food preparation like extrusion. Extrusion cooking of a blend of pea and cereals can result in a very good snack food having high content of digestible proteins with balanced amino acid profile and high fiber content. The present study was carried out to develop an extruded snack from a blend of rice and peas.

MATERIALAND METHODS

Garden peas of two cultivars (Pb-87 and Pb-88) were procured. Pods were shelled manually, grains were blanched and dehydrated. Other lot of peas of both the varieties was allowed to mature on the plant till the plant foliage dried up, harvested, pods were dried in shade for few days, shelled manually and grains were further dried in shade. Grits were prepared by grinding rice and pea (*dhal*) in a laboratory grinder to pass through 20 mesh sieve. Rice and pea grits were analyzed for crude protein (AACC, 1990), ash (AACC, 1990), starch (Ranganna, 1994) and amylose content (Juliano, 1971).

Blends were prepared by mixing rice and pea grits in the ratio of 100:0, 85:15 and 70:30. The blended samples were conditioned to 21 percent moisture by spraying calculated amount of water and mixing continuously in a laboratory mixer (Model N-50, Hobart, USA). Extrusion cooking of conditioned samples was carried out in a single screw extruder (Model 2003, C. W. Brabender, Hackensack, NJ USA) with 1.9 cm barrel diameter and 20:1 barrel length and diameter ratio. The extruder was fitted with a die nozzle having 3.75 mm diameter. The feed and compression and die zone temperature was maintained at 100°C, 130°C and 150°C, respectively. The extruder screw and feeding screw were run at a constant speed of 100 rpm.

Die pressure was measured using a pressure transducer (Dynisco Ltd. USA) just before the orifice. Readings were recorded every 10 sec for at least 2 min and average values were reported as kPa. Specific energy consumption (SEC) was calculated using expression described by Mason and Hoseney (1986).

SEC (Wh kg^{-1}) = (Amperes x Volts)/ (kg h^{-1} throughput)

The diameter of extrudates was measured using a Vernier calliper. Expansion ratio was calculated as the ratio of diameter of extrudates to the diameter of die nozzle. Density of extrudates was calculated as mass per unit volume using the expression:

Density (g/cc) = Mass (g)/ $\pi r^2 l$ (cc)

where $\pi = 22/7$ r = radius and l = length

WAI and WSI were determined by the method of Anderson *et al* (1969). The extrudates were ground to a mean particle size of approximately 180-250 μm. A 2.5 g sample was dispersed in 25 g of distilled water, using a glass rod to break up any lumps. After stirring for 30 min the dispersions were rinsed into tarred centrifuge tubes, made up to 32.5g and centrifuged at 5000 rpm for 10 min. The supernatant was decanted for determination of its solids content and the sediment was weighed. WAI and WSI were calculated using the expressions:

WAI (g/g) = weight of sediment/weight of dry solids

WSI (%) = weight of dissolved solids in supernatant x 100/weight of dry solids

The extrudates were fried in refined vegetable oil at 190°C for 1.5 min and evaluated for appearance/colour, texture, flavour/taste and overall acceptability by a semi-trained panel of six judges using nine point hedonic scale.

Data obtained were subjected to ANOVA using factorial completely randomized block design with statistical software CPCS 1 (PAU, Ludhiana).

RESULTS AND DISCUSSION

Table 1 shows crude protein, ash, starch and amylose content of rice and pea grits. It has been observed that pea grits were higher in protein, ash and amylose content and lower in starch content as compared to rice grits.

Table 1: Physico-chemical characteristics of rice and pea grits

Material			Crude protein (%)	Ash (%)	Starch (%)	Amylose (% of starch)
Rice grits			7.39	0.31	70.09	18.56
Pea grits	Pb-87	Mechanically dried	31.45	3.13	43.40	80.79
		Field dried	31.24	3.02	44.82	81.66
	Pb-88	Mechanically dried	32.64	3.63	46.51	75.67
		Field dried	32.05	3.56	46.89	74.73

Data with respect to the effect of cultivar, processing treatment and level of replacement of pea grits in rice grits on the extrusion behaviour and extrudate characteristics is presented in Table 2. Addition of pea grits resulted in a decrease in die pressure in all the blends. The feed material containing Pb-87 pea grits showed greater effect on die pressure than its counterparts containing Pb-88 pea grits. The die pressures observed in the present study were comparable with those reported earlier using same extruder under similar process conditions (Singh *et al*, 1996). Decrease in die pressure with increase in level of pea grits may be attributed to the increase in amylose content of feed material. Unlu and Faller (1998) observed a decrease in die pressure with increase in level of high amylose corn starch in feed.

Table 2: Mean values showing the effect of cultivar, processing treatment and level of replacement of pea grits in the rice grits on the extrusion behavior and exrudate characteristics

Factors		Die pressure (kPa)	SEC (W hr/Kg)	Expansion ratio	Density (gm/cc)	WAI (g/g)	WSI (%)
Cultivar	Pb-87	3420	342	2.06	0.267	5.20	27.23
	Pb-88	3688	315	2.03	0.277	5.01	25.46
Treatment	Mech. dried	3636	332	2.04	0.274	5.18	25.07
	Field dried	3471	324	2.06	0.271	5.03	27.61
Level	0	4104	364	2.15	0.265	5.45	30.87
(percent)	15	3496	325	2.03	0.273	5.02	26.23
	30	3061	296	1.96	0.278	4.85	21.92
CD (5%)	Cultivar (C)	96.56	4.96	0.016	0.024	0.09	0.50
	Treatment (Tr)	96.56	4.96	0.016	0.024	0.09	0.50
	Level (L)	118.27	6.07	0.019	0.030	0.12	0.62
	C×Tr	NS	NS	NS	NS	0.13	0.71
	C×L	167.26	8.59	0.028	0.042	0.17	0.87
	Tr×L	NS	8.59	NS	NS	0.17	0.87
	C×Tr×L	NS	NS	NS	NS	NS	NS

Addition of pea grits showed a significant effect on SEC. Addition of pea grits in feed material resulted in decrease in SEC, this decrease was more pronounced with Pb-88 pea grits. Decrease in SEC with increase in level of pea grits may be due to their higher amylose content. Unlu and Faller (1998) observed a decrease in torque with increase in level of high amylose corn starch in feed.

A decrease in expansion with increase in the level of pea grits in blend was observed. Decrease in expansion with increase in level of pea grits may be due to the higher protein and fiber and lower starch content of the feed material. Jones et al (2000) reported that an increase in protein and fiber content in feed material resulted in a decrease in expansion ratio. The extrudates having lower expansion showed higher density values and vice versa. An inverse relationship between expansion ratio and density of extrudates has been reported earlier (Singh, 1996).

Effect of replacement of rice grits with pea grits on WAI was found to be significant. A decrease in WAI with addition of pea grits may be due to the dilution of starch in rice pea blends. Jones *et al* (2000) reported a decrease in WAI with decrease in starch content. Colonna *et al* (1989) reported only little change in WAI for high amylose starches even with temperature up to 200°C. Level of pea grits significantly affected the WSI. Changes in WSI with incorporation of pea grits can be attributed to change in composition of the feed. Jones *et al* (2000) reported that fiber; starch and protein contents affect WSI. Bryant *et al* (2001) investigated that in comparison to waxy rice, high amylose rice exhibited lower WSI.

Table 3 shows the effect of replacement of rice grits with pea grits on the organoleptic characteristics of extruded snacks. Incorporation of pea grits showed a decrease in the colour/appearance score and an increase in the flavour, texture and overall acceptability score of extruded snacks. Replacement of rice grits with mechanically dried pea grits resulted in deterioration of colour score, but addition of field dried pea grits improved the score. Addition of pea grits in feed material showed significant effect on texture score. Addition of mechanically dried pea grits resulted in deterioration of texture score, but addition of field dried pea grits improved the score. Addition of pea grits resulted in an improvement in the flavour which was more pronounced with grits from field dried peas.

Replacement of rice grits with pea grits showed a significant effect on the overall acceptability score. The addition of mechanically dried pea grits resulted in deterioration of overall acceptability score, but the field dried pea grits improved the same. Pb-88 grits were found to be better with respect to the improvement of sensory quality of the snacks.

Table 3. Mean values showing the effect of cultivar, processing treatment and level of replacement of pea grits in the rice grits on the organoleptic characteristics (score out of 9.0) of extruded snacks

Factors		Colour/Appearance	Flavour	Texture	Overall acceptability
Cultivar	Pb-87	7.00	7.12	7.12	7.06
	Pb-88	7.08	7.09	7.26	7.14
Treatment	Mech. dried	6.48	6.81	6.75	6.65
	Field dried	7.60	7.40	7.64	7.55
Level	0	7.25	6.30	7.11	6.87
(percent)	15	7.09	7.40	7.28	7.25
	30	6.78	7.62	7.19	7.18
CD (5%)	Cultivar (C)	NS	NS	0.06	0.06
	Treatment (Tr)	0.09	0.07	0.06	0.06
	Level (L)	0.11	0.08	0.07	0.07
	C×Tr	0.12	0.10	NS	0.09
	C×L	NS	NS	0.10	NS
	Tr×L	0.15	0.12	0.10	0.11
	C×Tr×L	NS	0.17	NS	0.15

CONCLUSIONS

Incorporation of pea grits to rice grit resulted in a decrease in the energy consumption during extrusion cooking. Addition of pea grits resulted in a decrease in expansion ratio and an increase in density of extrudates. Incorporation of pea grits showed a decrease in the colour/appearance score and an increase in the flavour, texture and overall acceptability score of extruded snacks. Pb-88 grits were found to be better with respect to the improvement of sensory quality of the snacks. Replacement of rice grits with field dried pea grits resulted in improvement of acceptability score. Snacks obtained from a blend of rice and field dried pea grits in the ratio of 70:30 were the best.

REFERENCES

American Association of Cereal Chemists (1990) *Approved Methods of AACC*. St. Paul, MN, USA

Anderson R A, Conway H F M, Pfeiffer V F and Griffin E J (1969) Gelatinization of corn grits by roll and extrusion cooking. Cereal Science Today., 14: 4-12

Bryant R J, Kadan R S, Champagrte E T, Vinyard B T and Boykin D (2001) Functional and digestive characteristics of extruded rice flour. Cereal Chem., 78(2): 131-137.

Colonna P, Tayeb J and Mercier C (1989) Extrusion cooking of starch and starchy products. In: Mercier P, Linko P and Harper J M (ed) *Extrusion Cooking*. pp 247-319. AACC St. Paul MN, USA.

Jones D, Chinnaswamy R, Tan Y and Hanna M (2000) Physicochemical properties of ready-to-eat breakfast cereals. Cereal Food World, 45(4): 164-168.

Juliano B O (1971) A simplified assay for milled rice amylose. Cereal Sci. Today, 16: 334-341.

Mason W R and Hoseney R C (1986) Factor effecting the viscosity of extrusion cooked wheat starch. Cereal Chemistry, 63: 436-441.

Ranganna S (1994) *Handbook of Analysis and Quality Control for Fruit and Vegetable Products*. Tata McGraw-Hill Pub. Co. New Delhi, India.

Singh N, Singh B, Sandhu K S, Bawa A S and Sekhon K S (1996) Extrusion behavior of wheat, rice and potato blends. J Food Sci Technol, 31: 291-294

Unlu and Faller (1998) Formation of resistant starch by a twin screw extruder. Cereal Chemistry. 75(3):346-350.

SECTION - IV

POST HARVEST PLANT OPERATIONS AND QUALITY MANAGEMENT

16

Quality Management in Post-harvest Operation of Food Grains

S.M. Ilyas and R.K. Goyal

Abstract

India has become not only self sufficient in food grains but has also emerged as net exporter of wheat and rice. Acute famines in India have practically been eliminated, thanks to increased food availability at the national level, improved early warning systems and expanding infrastructure for more rapid transport of food grains and other products to needy areas. However, the existing post management system, strategies and policies don't seem to match new challenges. Despite government intervention in pricing and procurement of food grains, primarily wheat and paddy, India continues to suffer heavily on account of losses and wastages incurred during the post harvest stages, which are more manmade than natural. Some losses have already occurred before the arrival of food grains in the regulated and wholesale markets and continue till food grains reach their destination. What contribute to continuing losses and wastages are improper post harvest handling, government policies and their implementation, which even stifle growth and modernisation of grain markets and processors.

INTRODUCTION

India has become not only self sufficient in food grains but has also emerged as net exporter of wheat and rice. Per capita availability of rice and wheat per day increased from about 280 grams in 1967/78 to 450 grams in 2000/01. Acute famines in India have practically been eliminated, thanks to increased food availability at the national level, improved early warning systems and expanding infrastructure for more rapid transport of food grains and other products to needy areas. However, the existing post management system, strategies and policies don't seem to match new challenges.

Despite government intervention in pricing and procurement of food grains, primarily wheat and paddy, India continues to suffer heavily on account of losses and wastages incurred during the post harvest stages, which are more manmade than natural. Some losses have already occurred before the arrival of food grains in the regulated and wholesale markets and continue till food grains reach their destination.

What contribute to continuing losses and wastages are improper post harvest handling, government policies and their implementation, which even stifle growth and modernisation of grain markets and processors. The post harvest losses for food grains, according to the World Bank Study, are estimated to be 7 to 10 percent at the farm to market level and another 4 percent to 5 percent at the market and distribution level. For the system as a whole, the losses equal to 12 million to 16 million tonnes of grains -all grains – per year including 3-4 million tonnes of wheat and 5-7 million tonnes of rice. With an average per capita consumption of about 15 kg of food grains a month, these losses are enough to feed about 70 to 100 million people, about one-third of India's poor, or the states of Bihar and Haryana together, for about one year (Malhotra, 2002).

Indian agriculture is at a critical juncture. The changing character of agriculture especially of the food grains sector for instance is creating new demands on the post harvest management and marketing systems. This calls for proper post harvest handling and adjustments to food grain policies for procurement and distribution. Phase-wise technological advancement is shown in Table 1.The following post harvest operations in handling food grains will not only help farmers to fetch better price but it would also minimise losses occurring at various stages.

Table 1: Technology and phases of the post-harvest system for grain

Post-harvest Operations	Traditional Technologies	Intermediary Technologies	Industrial
Technologies			
Harvest	Manual	Manual and mechanised	Mechanised
Pre-drying	Standing or in shocks	In cribs or in shocks	
Storage in the ear	In traditional granaries	In cribs	
Threshing	Manual	Mechanised	Mechanised
Pre-cleaning		Mechanised	Mechanised
Drying	Natural	Artificial	Artificial
Cleaning and sorting	Winnowing in the wind	Mechanised	Mechanised
Storage in grains	In traditional granaries	In bags or in bulk	In bags or in bulk
Processing	Manual	Mechanised	Mechanised

Source: Lucia and Assennato, 1994

THRESHING AND SHELLING

These operations may be carried out in the field or on the farm, by hand or with the help of animals or machines. Depending on the influence of agronomic, economic and social factors, threshing or shelling is done in different ways:

Threshing/shelling

i. with simple tools
ii. with the help of animals or vehicles
iii. with simple machines operated manually
iv. with motorised equipment.

As observed above, combine-harvesters or picker-shellers can carry out the operations of harvesting and threshing or shelling simultaneously. Whatever the system used, it is very important that threshing or shelling be done with care. Otherwise, these operations can cause breakage of the grains or protective husks thus reducing the products quality and fostering subsequent losses from the action of insects and moulds. Transport of the product from the field to the threshing or shelling place must also be handled with special care, since it can bring about severe losses.

CLEANING

Cleaning should take place at the earliest opportunity in a food process both to prevent damage to subsequent processing equipment by stones, bone or metals, and to prevent time and money from being spent on processing contaminants, which are then discarded. In addition, the early removal of small quantities of food contaminated by microorganisms prevents the subsequent loss of the remaining bulk by microbial growth during storage or delays before processing. Cleaning is thus an effective method of reducing food wastage, improving the economics of processing and protecting the consumer. Types of contaminants found on raw food grains are mentioned in Table 2.

Table 2: Contaminants found on raw foods

Type of contaminant	Examples
Metals	Ferrous and non-ferrous metals, bolts, filings
Mineral	Soil, engine oil, grease, stones
Plant	Leaves, twinges, weed seeds, pods and skins
Animal	Hair, bone, excreta, blood, insects, larvae
Chemical [a]	Fertiliser, pesticides, herbicides
Microbial cells	Soft rots, fungal, growth, yeast's
Microbials products	Colours, flavours, toxins

[a]Not to be confused with adulterants (chemicals intentionally added to food which are forbidden by law) or additives (chemicals added to food to improve eating quantities or shelf life) (Source: Bernnan *et al.*, 1990).

It has been known for many years that the presence of dirt, dust and other foreign material in food grains accelerates deterioration because grain, which has not been cleaned, is more likely to heat than sound grain of the same moisture content. The reason for this has not been clearly established but it is obvious that extraneous material will harbour greater numbers of mould spores and bacteria, and in the case of insect, grass and dust will have higher moisture content at the same relative humidity. If the desired product is to be completely free of impurities and suitable for long-term storage, appropriate cleaning devices must be used, such as: winnowers, pre-cleaners and cleaner-separators.

Winnowers

These machines, some models of which may reach an output of one tonne per hour, can significantly contribute to improving product quality and marketing, especially in production areas. Sometimes operated by hand, more often motorised, winnowers are relatively simple machines that consist mainly of a hopper to receive the grain, a fan and a set of sieves.

Pre-cleaners

There are several models of pre-cleaner:

i. The circular pre-cleaner, in which impurities are eliminated principally by suction (output of about 150 quintals an hour)

ii. The drum pre-cleaner, in which the heaviest impurities are eliminated by passing the grain through the meshes of a drum, and the lightest are eliminated by suction (output of about 10 to 50 tonnes an hour)

iii. The fodder, in which the bulkiest impurities are eliminated by means of a coarse vibrating sieve.

Cleaner-separators

The usual recommendation is a set of sieves with oblong perforations for long grains, and for round grains, a clodder with round perforations and a sifter with oblong perforations.

SORTING

Like cleaning, sorting should be employed as early as possible to ensure a uniform product for subsequent processing. The four main physical properties used to sort foods are size, shape, weight and colour.

Shape and Size Sorting

The particle size distribution of a material is expressed as either the mass fraction of material that is retained on each sieve or the cumulative percentage of material retained. The shape of some foods is important in determining their suitability for processing or their retail value.

Size sorting (termed sieving or screening) is the separation of solids into two or more fractions on the basis of differences in size. It is particularly important when the food is to be heated or cooled as the rate of heat transfer is in part determined by the size of the individual pieces and variation in size would cause over-processing or under-processing. Additionally, foods that have a uniform size are said to be preferred by consumers. Screens with either fixed or variable apertures are used for size sorting. The screen may be stationary or, more commonly, rotating or vibrating.

Colour Sorting

Manual sorting by colour is still widely used but is increasingly expensive in labour costs, operator training and the space required for sorting tables. There has therefore been considerable development of machine vision sorting systems, which are said to have lower operating costs and greater accuracy than manual methods. These include monochrome (black and white), bichrome (4100 shades of red and green) and trichromatic or full colour (262 000 shades of red, green and blue, with optional infrared.

Small-particulate foods may be automatically sorted at high rates (up to 16 ton/h) using microprocessor-controlled colour sorting equipment. Particles are fed into the chute one at a time. The angle, shape and lining material of the chute are altered to control the velocity of the pieces as they pass a photo detector. The colour of the background and the type and intensity of the light used for illuminating the food (including infrared and ultraviolet options) are closely controlled for each product. Photo detector measures the reflected by a short blast of compressed air. The computer can store 100 named product configurations to enable rapid changeover to different products using an operator touch pad. Typical applications include peanuts, rice, and diced, carrot, maize kernels, cereals, snack foods and small fruits.

DRYING

It is essential that food grains be dried quickly and yet effectively. Food grains such as oilseeds, cereals and legumes differ in their biological make-up (having shells or outer coats and also inner kernels or seeds of different textures and compositions). Since each is processed in a particular way to meet the food habits of the local population, local drying methods vary. Drying affects the quality of grains and hence their value to purchasing organisations. For some grains and trades value is attached to appearance of grains, such as colour and texture, and for other grains and trades equal importance is attached to chemical composition, such as oil content and acidity.

Air is used as the drying medium and to conduct heat to the produce, causing water to vaporise: it is also used to convey the moisture vapour away from the drying produce. The moisture carrying capacity of the air is dependent upon the temperature and in fact, increases with the rise in temperature (e.g., at 30°C the air is capable of holding twice as much moisture as at 16°C).

There are two types of drying: natural and artificial. Within each type several systems may be considered. Different varieties of grains have specific drying characteristics and it is essential, therefore, to use a type of drying which will give the highest quality upon many factors, but principally upon the relative humidity of the air. The relative humidity of the air must be lower than the equilibrium relative humidity of the produce.

NATURAL DRYING

This type of drying utilizes a combination of sun and air but also requires time and some effort by man to spread and collect the produce.

In natural drying, care must be taken to avoid too rapid drying over-drying and to minimise excessive movement of the grains, which causes breakage or damage to the seed coat. Drying too rapidly causes certain seeds (e.g., pulses and spices) to swell and burst; with some types of products "case hardening" occurs, whereby the surface of the grains dries out rapidly, "sealing" the moisture within the inner layers. It also causes the grains to become wrinkled and scorched (or discoloured) in appearance; the greater part of the original carotene can be lost through oxidation and the riboflavin may be destroyed. Moreover, grains should not be exposed to rain and re-wetting during and after the drying period because stresses set up within the grain cause cracking. In natural drying care should be taken to eliminate dust and dirt, which represent extraneous material and accelerate and fungal contamination.

It should be noted that natural air-drying does not eliminate insects. Sun drying temperature in the tropics varies but it may be assumed that produce speared on matting in a layer not thicker than about 5 centimetres receives a drying temperature of about 36°C (higher temperature may be recorded, however) which is insufficient to kill insects present in the grains.

Artificial Drying

This involves the use of ambient temperature air and mechanical means of moving it through the produce, or air heated above ambient with or without the mechanical means of moving it through the produce. Air is heated by burning carbon and hydrocarbon fuels, utilising the heat supplied by internal combustion engines, and resistance heating by electricity.

In drying with direct-fired units fuel is burned and the hot combustion gases pass through the produce or are collected by the air stream from a fan and distributed throughout the produce being dried; with this system the produce can become contaminated by unburnt fuel, fumes and soot. Contamination of the product does not occur in indirect-fired driers where hot gases pass into a heat exchanger around which the drying air circulates and picks up heat before it is distributed through the produce.

In drying crops in deep layers, excessive temperature increase at constant air moisture results in drying grain to an undesirable low level of moisture content at the bottom of the layer before moisture content in the grain at the surface is near the desired level. The slow, steady rate of drying achieved by this method can yield a high quality produce for some crops such as rice.

In drying grain in shallow layers the maximum permissible drying temperatures recommended depend upon the use for which the grain is intended; examples are given in Table 3. These are broadly recognised values not appropriate to all areas or varieties. Experience may indicate that temperatures in excess of these may be used satisfactorily. Initially, however, the temperatures are always recommended when quality is a factor.

Table 3: Thin layer drier operating temperature conditions

Produce and intended use	Maximum recommended drying temp., °C
Grain to be used for livestock feed	74
Grain for human consumption	57
Grain for milling and manufacturing	60
Seed grain or brewery grain	43
Rice for human consumption	43
Beans for human consumption	35

(*Source*: Hall, 1980)

In order not to destroy grain viability, the maximum heating time air at different temperature is limited according to the moisture content of the grain. Pedersen (1962b) has given relevant data as recorded in Table 4. "Stage" drying is used with some commodities to maintain certain quality criteria, particularly with rice. Instructions for rice, which is harvested at 25- 27% moisture content, are: (i) dry to approximately 20%, (ii) cool to atmospheric temperature, (iii) store for approximately 24hrs, (iv) then repeat the drying and storage sequence for at least two more stages. In the case of rice, cracking is said to be kept to a minimum by this procedure.

Table 4: Relation between moisture content, time and permissible heating of the drying air

Heating time	Grain moisture content, percent			
	15	20	25	30
Minutes		Degree C		
15	59	55	52	50
30	56	52	49	47
60	53	49	46	44
120	50	46	43	
480	44	40		
1440	39			

Source: Pederson, T. T 1962b

Excessive drying due to the drying air being at a low relative humidity need not occur in continuous flow driers. With batch driers, the final moisture gradient through the batch is likely to be somewhat less when working with high rather than low moisture content air at the same hot air temperature.

STORAGE OF FOOD GRAINS

The clean, dry and cool grain may be stored either in bags or in bulk and typical flow chart for storability assessment of food grains is shown in Fig 1.

Bag Storage

The food grain may be stored in clean and sound jute bags. Old jute bags if used should be thoroughly checked for holes and insect infestation. Torn jute bags should be properly stitched before filling the grain. To check infestation in old jute bags, these may be sprayed with 0.5% solution of Malathion, 50% EC in water and should be dried before use. The grain bags should be stored in pucca room by using wooden crates or planks. As a prophylactic measure, the walls and floors of empty room may be sprayed with 0.5% Malathion, 50% EC at the rate of 3 litres emulsion per 100 sq. meter of surface area.

Cap Storage

It is popularly known as cover and plinth storage system. This system is often used for short-term storage in dry areas. It consists of stacking bags pyramids on platforms that can be protected in case of bad weather. The platforms on which the bags are stacked must effectively protect the grain against termite attacks. Covered by concrete or tar or made up of a layer of building blocks covered with tarpaulins or plastic, the platforms must prevent the exposure to rising damp. To achieve this end, not only must the sites of the storage area be carefully chosen but also drainage ditch for rain-water runoff must be dug around the platforms. To keep rainwater from failing on the grain, it is important to cover the pyramids of bags with tarpaulins.

Warehouses and storehouses

Whether simple huts that have turned into storehouses or modern well-equipped warehouses, storage buildings must meet the following requirements:

i. Prevent the grain from getting wet

ii. Protect the grain from high temperatures

iii. Prevent the access of insects, rodents and birds to the storage places

iv. Facilitate monitoring of grain conversation

v. Permit timely insecticide treatment used to move and transport the bags.

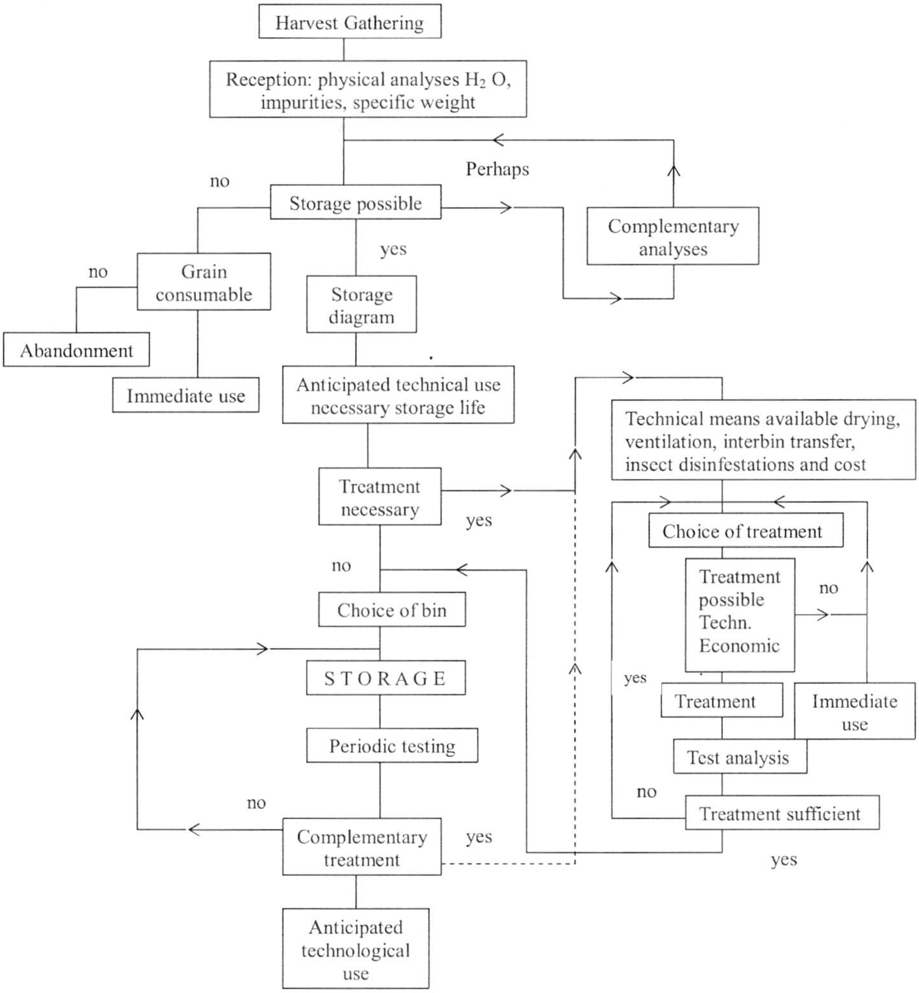

Fig. 1: Storability assessment flow chart

The effects and consequences of mould activities on stored grains are shown in Table 5.

Table 5: Effects and consequences of mould activities on stored grains

S. No.	Process	Effects	Consequences
1.	Adverse quality changes	Dull appearance Musty odours Visible moulds Reduced germination Germ damage, discolouration Increased free fatty acids	Possible degrading Rejection for malting Degrading Rejection of operations
2.	Aggregation of product	Closing of pipes, augers Sticking to bin walls Bridging of bin contents Aggregation and/or fusion of bin contents	Interruption of operations Uneven pressure effects, partial bin collapse. Dangerous air space Cleaning out costs, unusable facilities.
3.	Heating of product	Bin-burning	Damage to product and premises, Possible degrading, rejection, extra costs, Could lead to fire-burning explosions
4.	Contamination of product by harmful substances	Mycotoxins Respiratory/ allergenic effects	Livestock poisoning, feed Loss of markets Chronic human health problems Breathing problems in animals and humans Employees handling grain may need to be resigned.

Bulk Storage

The grain may be stored in metal bins of different capacity or in pucca kothi made of cement and bricks. Other improved traditional storage structure like kacchi kothi and kuthla may also be used. Metal bins designed and developed by Indian Grain Storage Institute, Hapur, and manufactured and marketed by State Agro-Industries Corporations have become very popular among farmers for storage of food grains. The Indian Grain Storage Institute, Hapur has developed 31 designs of modern farm level storage structures of capacities ranging from 0.09 to 14.5 M.T. to suit the requirement of farmers in different regions. These have been found suitable for storage of different commodities of food grains like wheat, paddy, maize, pulses, sorghum and seed grain, provided the moisture content of the grain at the time of storage is kept within the safe storage limit. The improvement have been carried out in traditional storage structures made of timber, bamboo, paddy straw, mud drum, etc. which are used in different parts of the country. Table 6 shows the safe moisture level for storage of food grain at 26-27°C.

Table 6: Safe moisture content of food grains for storing at 26-27°C

Food Grains	Moisture Content, %
Maize (yellow)	12.0 – 13.5
Maize (white)	11.0 – 11.5
Paddy	12.0 – 13.0
Rice	11.0 – 12.0
Other course grains	10.0 – 12.0
Sorghum (Jowar)	12.0 – 13.5
Wheat	12.0 – 14.0
Wheat flour	10.0 – 12.0
Pulse	12.0 – 14.0
Lentil	13.0
Groundnut (with shell)	10.0
Groundnut (Grain/shelled)	8.0
Soybean	10.0 – 11.0

Source: Ahmed, 2001

Low capacity silos for farm storage

There are several types of traditional storage structures, each adapted to the climate of the country. Their common feature is the use of locally available materials. Some examples are the enclosed earthen granaries of the dry zones and the ventilated granaries made of plant fibre and wood that are used in humid zones. In dry zones, the risks of stock degradation come mainly from insects and rodents, and they are generally lower than the risks in humid zones, where stocks are attached not only by these pests but also by moulds.

There are two possible approaches to lowering losses from pest attacks: improvements to traditional storage structures, and new structures built from non-traditional materials. The first approach has produced improvement in the construction of earthen granaries (mixing small quantities of cement with the earth or careful finishing or smoothing of the silo walls). Other more innovative experiments involve building small-capacity (1 to 2 tonnes) silos with bricks or breezeblocks made of reinforced earth, or with mud or unbaked bricks. The second approach has introduced new storage structures that generally require non-traditional materials and construction techniques. Examples are silos made of concrete, reinforced concrete, or metal.

Square or rectangular bins are generally flat-bottomed. They require a higher per quintal investment but make the best use of the available sites. In order to avoid the disadvantages of a potential rise in temperature, and to guarantee good storage, storage bins are often equipped with ventilation systems backed up by temperature controls.

PACKAGING

Food grain packaging over the years has only been limited to materials such as jute sacks. However, there are alternative packaging materials that are technically suitable for packaging food grain. In case of food grain, the requirement for packaging material should be:

- To prevent contamination of the packed edible items from the packaging material (or its constituents)
- Moisture resistance
- Rot resistance
- Good shelf life irrespective of exposed or covered storage
- Breathability

Other factors deciding the choice of the packaging materials are:

- Better-cost economics
- Easy availability of the packaging material
- Reusability to improve overall cost economics
- Subsidy borne by the government

Deterioration and losses of products, during transport and storage, depend upon a series of physical, chemical, biological and human factors. Proper packaging is an important element in reducing losses, especially in the tropics; climate considerably increases the risks of grain deterioration.

There are various types of packaging for agricultural commodities, appropriate to the products characteristics and to the marketing system. Woven bags made of plant or plastic fibres are the usual types used for grain. Rather reasonably priced bags, which still fulfil the functions described above, can normally be made from such fibres. The choice of the type of bag should take into account not only its inherent toughness and resistance to humidity, sunlight and pests but also the type of handling anticipated.

TRANSPORT

Transport generally marks the passage from one stage of the post-harvest system to the next. Transport, whether traditional or mechanised, is needed to move the agricultural commodities:

i. from the harvest fields, to the threshing or drying site
ii. from there to the farmers storehouses or to collection-centre warehouses
iii. from there to the processing industries or to bigger central storage buildings (often much farther from where the commodity is grown)

iv. from these industries or storage buildings to wholesalers or retailers for final marketing.

In many parts of the world, farmers and peasants live far away from any road on which grain can be transported to places of collection, storage, or marketing. Therefore, products are often carried in small quantities over very bad roads or footpaths. This entails long transport times and high costs per unit of product transported, leading to a substantial decrease in incomes. This does not encourage farmers to increase their production.

Indeed, increasing production requires small farmers not only to develop storage buildings but also to read just the local road network to the product transport. To alleviate these disadvantages, the road network would have to be improved to permit development of a small-scale transport system to meet the needs of the most distant production areas.

In places where the road network is undeveloped and agriculture is traditional, people, donkeys, camels, and sometimes horses generally transport products: oxen are more often used as draught animals. Otherwise, pickup trucks, buses and taxis are often used to transport products to the collection, storage and marketing places.

QUALITY ASSURANCE, CODES AND STANDARDS

Small and medium sized food processing businesses all over the world increasingly have to consider the production of good quality products as essential to their survival. Consumers and buyers are becoming more aware about the importance of safe, high quality products. There is now a premium on quality.

Food quality

Food quality, as distinct from food safety is the extent to which all the established requirements relating to the characteristics of a food are met. Common examples of quality characteristics of food, excluding the food safety characteristics (Ali, 2003) are:

- Identity of a food in relation to a standard (e.g., standardized food)
- Declared gross or net quantity (e.g., weight or volume) of a unit of food or net fill of a food container
- Appearance (e.g., size, shape, colour)
- Flavour
- Aroma
- Texture

- Viscosity
- Shelf-life stability
- Fitness for use As human food
- Wholesomeness
- Adulteration
- Packaging
- Labelling

Quality assurance systems

Quality assurance systems in the food industry are much more extensive in scope than quality control programs. They include the inspection, testing and monitoring activities of quality control programs, along with additional activities that are devoted to prevention of food safety hazards and quality defects. An example of quality assurance system is the ISO 9001:1994 quality assurance system standard, which was replaced by the ISO 9001:2000 quality management system standard.

ISO 9000 quality system standards

These quality system standards, which were introduced in 1987, were revised in 1994 and again in 2000. In the 2000 revision, quality management principles and fundamentals of quality management systems were formally recognized and incorporated into the ISO 9000 family of standards. The 2000 revision, which was the first major revision of the standards, resulted in following three standards:

- ISO 9000:2000 Quality management systems- fundamentals and vocabulary
- ISO 9001:2000 Quality management systems- requirements
- ISO 9004:2000 Quality management systems- guidelines for performing improvements

The two standards ISO 9001:2000 and ISO 9004:2000 have been referred to as a constant pair, indicating that they are compatible with each other.

GMPs and HACCP

The good manufacturing practices (GMPs) are used for monitoring the safety of consumer foods and for inspection of establishments that process, package, handle and store foods. The emergence of HACCP system for achieving food safety has reinforced the importance of GMPs to the extent that the prerequisite programmes are considered as the necessary foundation on which an effective HACCP system is built.

The universally recognized Seven Principles of HACCP are:

- Principle 1: Conduct hazard analysis
- Principle 2 : Determine critical control points
- Principle 3 : Establish critical limits
- Principle 4 : Establish monitoring procedures
- Principle 5 : Establish corrective action procedures
- Principle 6 : Establish verification procedures
- Principle 7 : Establish record-keeping and documentation procedures

As per the requirement of the domestic as well as global market, the processed food should be well manufactured by observing HACCP and good manufacturing practices (GMP) so that, the product is of very high quality and at most competitive price. This calls for a complete paradigm shift in the production and post-production sectors.

Grains are used as feed for animals, as food for humans and as feedstock for industrial processes. As a feed, the nutritive properties have to be healthy. As a food or feedstock the milling and processing qualities of grains have to be acceptable. The desired attributes differ greatly for the various grains, as in shown in Table 7.

Table 7: Grain quality attributes of importance for the milling processing or human consumption of different grains

Maize	Rice	Soybeans	Wheat
Hardness	Milling quality	Splits	Hardness
Kernel weight	Cooking quality	Kernel weight	Kernel weight
Fat acidity	Kernel colour	Protein content	Gluten content
Soluble protein	Eating quality	Oil content	Fat acidity
Turbidity		Free fatty acids	Flour yield
Sedimentation		Iodine number	Semolina yield
Amylose		Phosphatides	Ash content
		Oxidation value	Dough quality

Brooker, D.B., Bakker-Arkema, and F. W., Hall (1992). Gordon, B. and William, C. (1994).

The properties of grains that determine the overall quality of a cereal, oilseed or pulse include

1. Appropriately low, uniform moisture content
2. High test weight
3. Low percentage of foreign material

4. Low percentage of discoloured, broken heat, damaged and shrunken kernels
5. Low breakage susceptibility
6. High milling quality
7. High oil content and recovery
8. High protein content
9. High nutritive value
10. High viability
11. Low mold count
12. Low carcinogen content
13. Low insect damage

CONCLUSION

The food post-production sector is a vital component of the food system, and its development can stimulate improvements in the creation and use of resources. One indication of the need for increased allocation of resources for research and development in the post-production system is the high level of post harvest losses. Losses are attributed to a combination of factors affecting the way the crop is grown, harvested, cleaned, handled, dried, stored, milled, transported and marketed. These losses are either outright physical losses, or deterioration of quality, which reduces the commercial value. The big gap has been in the development of technologies for post harvest handling suitable to the conditions prevailing in the world in general and India in particular and in the institutional organisations to enable local farmers and processors to use technology to improve productivity in their operations. The high quality of raw materials has attained much importance due to tough competition under liberal economy on one hand and need for quality raw produce for further processing. The unfortunate scenario of "no premium for the quality" is changing surely and fast.

Besides, primary processing, which includes handling operation at each stage enables the creation of agro-processing industries to produce value-added products, assuring greater financial returns and generating employment opportunities, thus realising its great potential in reviving rural economy. The requirements of primary processing are varied and they vary with the commodity and are location specific involving different sets of unit operations. These unit operations also vary with crop to crop, commodity to commodity. For example, the typical unit operation for food crops are harvesting, cleaning, grading, drying, curing, shelling, decorticating, parboiling, dehusked, polishing, size reduction, expelling, mixing, blending, fortification, packaging, quality control, storage, transport, marketing and by-product processing. All the relevant codes and standards including BIS,

PFA, ISO, Codex Alimentarius need to be properly understood and implemented. In this the farmer/producer requires help from all quarters viz., research institutions, government departments and all other concerned. Some enlightened and dedicated NGOs could be very helpful in spreading the message of quality at all levels.

The experience of last three decades has shown that post harvest technology has potential to check losses and to create rural industries for employment generation and minimisation of role of middleman. It has been well established that the process of industrialisation has shifted the food, feed and fibre industries to urban areas. This process has resulted in capital drain from rural to urban areas, decreased employment opportunities in the rural areas, balance of trade in favour of urban sector and mis-matched growth in economy and standard of living, widening gap between rural and urban people. It is possible to evolve appropriate technology, which can establish agriculture based rural industries.

The farmer, whose role is reduced to supplier of raw material only, can be transformed into producer-cum-processor thus getting more dividends for hard labour put in and risk taken and generating resources for socio-economic advancement keeping pace with the modern times. This would also enable him to plough back a part of his income in form of bigger and better inputs for higher productivity turning a vicious circle into a virtuous one. The face of Indian Agriculture from a primitive profession of producer and supplier of raw material needs to be transformed into dynamic one which is run on commercial lines and sells preserved/processed produce of high quality (best in the World) at competitive price. But the transformation has to be quick- the competition will not allow it otherwise.

REFERENCES

Ahmed, S.G. (2001) Safe food grains to Fight Hunger, Intensive Agriculture, September- October, pp. 34-37.

Ali, I. (2003) Food Quality Assurance-Principles and Practices. CRC Press, Florida

Brennan, J.G., Butters, J.R., Cowell, N.D. and Lilley, A.E.V. (1990) Food Engg. Operations. Elsevier Applied Science, London, (a) pp. 69, (b) pp. 121-122.

Brooker, D.B., Bakker-Arkema, F.W. and Hall, C.W. (1992) Drying and Storage of Grains and Oilseeds, Van Nostrand Reinhold, New York.

Gordon, B. and William, C. Eds. (1994) Primary Cereal Processing. VCH Publishers, New York.

Hall, D.W. (1980) Handling and Storage of Food Grains in Tropical and Subtropical Areas, FAO Plant Production and Protection Series No. 19, FAO, Rome.

Lucia, M. and Assennato, D. (1994) Agri Engg. in Development: Post Harvest operations and Management of Food grains. FAO Agri. Services Bulletin 93, FAO.

Malhotra, S.P. (2002) Post Harvest Management, India Grains, March pp 3-6.

Pedersen, T.T. (1962) Technical development in grain harvesting and storage problems related to combining in humid areas. Proc. UN Conference on the Appl. of Sci and Tech for the Benefit of Less Developed Areas, Geneva, Feb, pp. 5.

17

Strategies for Upgrading Quality of Food Grains at Competitive Cost During Storage and Handling

Rajinder Singh

Abstract

'Made in India' is no more a sign of poor quality. However, in the field of food grain exports, India is still lacking the required determination. The Indian food grains get rejected in the International market because of its out-dated post harvest technology (PHT). The Food Corporation of India (FCI) is the nodal national agency for the procurement, storage, processing and transporting of food grains, and their further distribution through the Public Distribution System (PDS). The government fixes the procurement price as well as the distribution price of the food grains, and the FCI has to accept it. The government compensates the resulting loss through subsidy. The government, through FCI, must encourage cultivation and marketing of premium quality food grains, by offering suitable incentives to farmers. The cost of marketing, storage and transporting of food grains can be considerably reduced with simultaneous improvement in quality of the end product, if the prevailing bag handling is replaced by bulk handling, storage, and transportation. However this needs closer coordination between various departments of the Ministry of Food and Ministry of Railways. The President of India revealed his vision to make India a developed country by the year 2020. To achieve this, the current post harvest losses to the tune of 10 percent must be reduced to 2 percent. This is possible if the government takes bold political decision supported by equally firm and time-bound necessary

administrative steps for implementing the adoption of this modern system at the national level. This way, India can assure world-class quality of food grains at competitive prices and establish its name in the world market, as it has recently done in the case of automobiles, textiles, pharmaceuticals, and leather goods.

INTRODUCTION

Production of food grains is on seasonal basis whereas its demand is fairly stable throughout the year. Wheat and paddy constitute the major components of food grains. The production and marketable surplus of wheat and paddy have steadily increased as a result of the minimum support price, which encouraged the farmers to adopt the modern tools and technology. On an average 46.6% of wheat and 24.5% of paddy production, is retained by the farmers for their own use and the remaining portion is available for sale in the open market. As far as storing the grains for his domestic consumption is concerned, the farmer adopts the traditional or modified traditional storage structures or even the modern steel bin. He is fully conversant with this technology and can reduce the grain losses to negligible limits. However for commercial grain storage, he has little experience or knowledge and ends up with huge losses.

Post Harvest Options for Farmers

After harvesting, in general the farmer as an individual has four options

1. He can retain a portion of the produce for his own domestic consumption (his own family and for his permanent labour, etc.), and immediately sell the remaining portion to get ready cash for buying inputs for the next crop and meet the day-to-day expenses.
2. He can retain a portion of the produce for his own domestic consumption as well as for sale at a later date (to take benefit of higher off-season prices) and sell the remaining portion immediately.
3. He can retain the entire produce (depending on his resources) for his own consumption as well as for deferred sale, as raw produce.
4. He can retain the entire produce (depending on his resources) for his own consumption as well as for deferred sale after processing, as a value added product.

Invariably, the farmers opt for choice number one (which is easiest but gives minimum turnover per unit weight), because they lack resources, which are required for other options. The small, medium and even some big farmers need ready cash to buy inputs for the next crop and to pay for other day-to-day expenses. They lack appropriate marketing knowledge and the associated facilities / experience for storing grains scientifically and economically on commercial basis, without incurring huge grain losses.

Moreover, huge fluctuations in the price structure also discourage him from storing grains beyond his domestic needs. Only the very big farmers have the required resources of their own. But even they go for the second option and hesitate to adopt the third or fourth options (which give maximum turnover per unit of weight), because they do not have access to the marketing / technical professionals and there is no insurance cover to protect them against unforeseen risks, as such they are not ready to take the risk involved in the commercial storage and processing.

Contract Farming

The farmers are not leaving wheat-paddy rotation because this rotation is comparatively free of risks and gives assured maximum annual returns per unit of land. This practice has increased the demand for irrigation water, which is dangerously lowering the water table. Now this has become a major cause for the worry. For the past one and a half decade the agricultural experts and planners have been advocating diversification of crops, but so far, at the best, it has met with a very limited success.

The Punjab government introduced a bold and innovative scheme for multi-crop multi-year contract farming to encourage farmers to diversify Punjab's agriculture (Anonymous 2003). Under this scheme, four to twenty five thousand acres of cropland is to be diversified from wheat-paddy rotation to other crops during 2003 to 2007 as shown in Table 1. To make this scheme a success, the nodal agency, should

i. Provide seeds of high yielding varieties from reputed companies,

ii. Facilitate technical supervision and follow up on agronomic practices,

iii. Buy back entire produce on a market linked bonus purchase plan.

However, the success of the scheme will depend greatly on the sincerity of the implementing agencies and the rules and regulations under which they may have to operate.

Table 1: Contract Farming Plan for the Years 2003 to 2007

Sr. No.	Crop	Target Plan in Acres for Years				
		2003	2004	2005	2006	2007
1	Hyola	75,000	2,00,000	3,00,000	4,00,000	5,00,000
2	Barley	5,000	15,000	25,000	40,000	70,000
3	Winter Maize	3,000	5,000	11,000	12,500	15,000
4	Durum Wheat	1,00,000	2,00,000	3,00,000	4,00,000	5,00,000
5	Sunflower	12,500	40,000	1,00,000	1,50,000	2,25,000
6	Spring Corn	5,000	15,000	25,000	40,000	80,000
7	Basmati	25,000	50,000	70,000	1,00,000	1,50,000
8	Kharif Corn	1,50,000	3,00,000	4,00,000	5,00,000	7,00,000
9	Guar Gum	5,000	6,000	7,500	10,000	14,500
10	Castor/Jatropha	5,000	10,000	20,000	40,000	50,000
11	Groundnut	1,000	1,5000	15,000	20,000	25,000
12	Org. Basmati	1,00	3,000	5,000	7,000	15,000
13	Vegetables	2,000	4,000	6,000	10,000	20,000
14	Others	10,500	50,500	25,500	70,500	1,35,500
	Total	4,00,000	9,00,000	13,00,000	18,00,000	25,00,000

Handling-Care of Grains at Farmer's Level

It is observed that, the farmers properly clean, dry and store the grains meant for their own domestic consumption. But the grain meant for immediate sale is neither properly cleaned nor dried before selling. Rather sometimes, some unscrupulous farmers in connivance with the officials of the purchase agencies intentionally add extra moisture to increase weight. Quality, of grains and for that matter of any product, depends upon the urge and motivation of the producer. In a free and competitive society the quality in general is related to the profit margins.

If the farmer is assured of a bonus price (commensurate with the extra effort put in by the farmer) for premium quality grains, he will do his best to bring clean, dry and good quality grains into the market. This will also reduce congestion in the markets because cleaning, which is the most time consuming operation in marketing, will not be needed.

Further, if the farmer himself becomes the processor and wants to market value added processed products in the local or export markets, he will become highly quality conscious and to promote his trade name he will maintain the quality of his raw material also. In fact, to achieve this, all he needs is a package of adequate incentives for proper motivation, technical/marketing knowledge, and insurance cover against the unforeseen risks/failures.

Prevailing Commercial Storage

The commercial storage of foodgrains is done in bag form in CAPS (Cover and Plinth storage) or warehouses, and, in bulk form in RCC or steel silos depending on the facilities available. Foodgrains stored in bags are relatively more susceptible to damage by pests. Under most suitable dry conditions grain has been held in bags, in India, in satisfactory condition for periods up to 2 years. In coastal areas deterioration is faster and grain can be held for periods varying between 8-12 months. On the other hand same grain can be held in bulk practically without any damage for periods up to 5 years irrespective of location (Pingale, 1976). The apparent advantages of storing in bulk are:

a. Greater storage capacity per unit volume of space,

b. Mechanized loading and unloading is easier and cheaper

c. No cost of purchasing gunny bags and dunnage

d. No danger of rodents

e. Insect infestation is considerably less

f. No or minimum use of insecticides

g. Minimal effect of outside environment on foodgrains

The storage system that needs minimum use of insecticides (sprays or fumigation) should be preferred. In bulk storage, infestation is lesser due to conditioning of grains and considerably smaller surface area in contact with external environment minimizing the effect of changes in air temperature and humidity.

The inter-granular space of lower grain layers probably has high concentration of carbon dioxide, especially when the bulk is large enough to create proper independent ecosystem of its own (Anon, 1998). Under this ecosystem insects feeding on upper layers get lesser surface area while those which are deep feeders have to encounter higher concentration of carbon dioxide and lesser oxygen which restricts their population growth. The bag storage does not have such advantages but some other advantages as:

a. Small lots of a number of crops or varieties can be stored in the same space

b. Infested bags can be easily segregated and fumigated.

c. Each bag can be handled independently

Commercial Storage Facilities in Public Sector

There are a large number of small farmers who lack capacity to hold grains and are forced to sell their produce immediately after harvest. The Central and State Warehousing Corporations were established on 2nd March 1957 for the creation of a negotiable paper – Warehouse Receipt- to provide institutional credit facilities to the farm sector and storage facilities to increase their holding power (Maseeh, 1990). Under this system the farmer can store his grains in the warehouse and get a receipt against which he can purchase the field inputs for his next crop. He can sell his crop to the warehouse or in the open market at a later date if he so chooses.

Food Corporation of India was established on 1st January 1965 to implement government policies with regard to food procurement, transport, storage, and distribution. The Central Warehousing Corporation provides warehousing facilities at centres of All India Importance. The State Warehousing Corporations provide similar facilities at State, District, Block and Village levels. The storage facilities are mainly for bag storage though bulk storage is also provided to a limited extent. In CAPS as well as warehouses the commodities should be stacked on a moisture proof floor cover, such as polythene sheeting, or on dunnage such as wooden pallets or platforms or crates. The commercial storage by the FCI and other government agencies is being done mostly in bag form only. Negligible amount (2%) is stored in bulk form (Table 2).

Table 2: Bulk Storage Installations of FCI (Capacity in tonnes)

	Location	Tall vertical silos	Low height flat storage bins (for bulk storage)
1.	Hapur (U.P.)	10,000	5,500
2.	Calcutta (W.Bengal)	19,200	—
3.	Kanpur (U.P.)	—	72,000
4.	Manmad (Maharashtra)	—	84,000
5.	Borovli (Maharashtra)	—	1,04,000
6.	Gaya (Bihar)	—	32,000
7.	Faridabad (Haryana)	20,000	—
8.	Naraina (Delhi)	20,000	—
9.	Khurja (U.P.)	20,000	—
10.	Lucknow (U.P.)	20,000	—
11.	Mandi Gobindgarh (Pb)	20,000	—
12.	Jagraon (Punjab)	20,000	—
13.	Moga (Punjab)	20,000	—
	Total	1,69,200	2,97,500
	Grand Total	4,66,700	

Open Storage of Foodgrains under Cover and Plinth (CAP) System

Foodgrains (wheat and paddy only) are stored in open CAP system. Only 5 percent of the Punjab wheat is stored in godowns, whereas 95 percent is lying in open plinths resulting in 8-10 percent post harvest losses (Anonymous, 2002 A). The grain bags are stored on wooden crates in the form of standard pyramid shaped structures on plinths made of brick and cement. Bags are stacked in (10+6) arrangement so that one layer contains 96 bags. At the top of 15 layers a dome shape is made with about 60-100 bags, so that one plinth of CAP holds 1550 bags. The foodgrains are covered with specially fabricated black polyethylene covers of 1000 gauge LDPE films to store 150-155 tons of foodgrains in a space of 9.5 x 6.4 x 5.2 meters.

The expected life of the film is more than one year and on an average 0.8 ton plastic is used for every 1000 tons of foodgrain stored per annum. Unit cost wise, CAP storage is almost one seventh of the cost of storage by conventional godowns (Srivastava et al, 1989). LDPE CAPs are more economical than PVC CAPs. Due to paucity of covered storage space, this type of storage system is resorted to in major wheat and paddy procurement areas, particularly in Punjab, Haryana, Rajasthan, Uttar Pradesh, Madhya Pradesh, Andhra Pradesh, and Maharashtra.

This is a temporary storage system wherein foodgrains like wheat, paddy, maize, gram, and Jawar may be stored for a short duration (Approximately six months) and is resorted to only when adequate covered godowns are not available. However this is not followed in practice and the bags are stored in CAPs for even two or three years, which is unscientific. The FCI at present has a total covered storage capacity of about 3 million tonnes under the CAP system (Anonymous 1998). The scientific storage (CAP or Warehouse or Silo system, if stored within the permissible time limits) would lead to decrease in loss of foodgrains, improve qualitative preservation, and ultimately realize better prices.

Management of CAPs

A code of practice for storage of foodgrains in CAP storage has been developed which involves timely prophylactic and curative treatments with recommended insecticides and fumigants. Adequate rodent control measures also have to be adopted and time-to-time aeration is required during sunny days to reduce moisture content of the grains. Anti termite treatments with suitable insecticides is very essential in CAP storage, otherwise wooden crates and gunnies may be completely damaged by termites leading to collapse of the stacks (Anonymous, 1998).

Sites selected for CAP storage must be a high plinth area preferably with pucca masonry work. It should have a proper drainage system, adequate security measures, absence of wild vegetative growth, and good approach roads. This type of storage needs constant vigilance and regular monitoring of quality.

The main advantage of this system is that it is comparatively cheap and serves the purpose of scientific storage for short periods under dry weather conditions. However the system has number of disadvantages.

Warehouses

The foodgrains are stored in bags as in the case of CAPs. However, in case of warehouses the stacks should be built clear of all walls, buttresses, roof supports, roof trusses and roof itself to allow access all round and to permit free circulation of air. It should be possible for a person to pass easily around all sides and over the top of each stack. Standard designs of warehouses are available for capacities varying from about 100 tonnes to 5000 tones.

The main advantage of warehouse storage is that the bags can be individually identified, stored and handled. In case of infestation the bags can be identified and segregated.

The major disadvantages of warehouses are: the space utilization is poor, the handling cost is very high, different varieties and commodities are subjected to only one set of environmental conditions, in-spite of all precautions the rats can enter the warehouse (particularly if they are already inside the bags), the inside environment gets easily affected by the changes in the outside environment, the end quality is not very good and the overall storage cost (including cost of land, structures, dunnage, bags, labour, insecticides, damaged grains, and over-head charges) is more than in the case of modern silo system..

The maximum safe storage period for warehouses is only one year and that for CAPS it is only 6 months under ideal conditions, but in practice grain is stored in CAPS even up to 2 to 3 years till it becomes unfit for eating even by the animals.

Bulk Storage and Handling

Bulk storage and handling of foodgrains is the accepted practice in almost all the developed countries of the world. Foodgrains can be stored in bulk form in individual bins, or silos/elevators made of steel or concrete. Even the existing warehouses can be easily modified to store grains in bulk. Precaution should be taken to ensure that the height at the walls is not more than 1.5 meters and grain is allowed to maintain the angle of repose so that the height at the centre is from 4.5 to 5.0 meters. The capacity of the godown increases by 30%, which reduces the unit cost of storage. A silo or elevator is one or more bins with facilities for loading, cleaning, drying, bulk storage, regular monitoring of grain quality, aeration / conditioning, unloading, weighing and bagging of grains.

The individual bin can be hexagonal or circular or star shaped in cross section. A typical battery of reinforced cement concrete-cum-brick hexagonal bins, of total storage capacity about 2030 tons, consists of 58 (12 x 3 + 11 x 2 = 58) bins of 35 tons capacity each. The bins are arranged in five rows, in such a way that three rows are of 12 bins and two of 11 bins, with adjacent sides common. This pattern allows a minimum surface area to be exposed to the outside environment. These bins are manually filled from the top inlet, and unloaded into bags from the raised hopper bottom, by operating a lockable valve.

The circular bins are most common. These can be made of galvanized sheets or reinforced cement concrete. The RCC bins generally have common walls and space between the silos is also utilized for storage (star shaped bins). Silos are cylindrical with hopper bottoms. Such large sized silo systems (RCC/GI) can be easily integrated into a fully mechanized grain marketing, storage and processing facility, by adding a flour mill and a rice mill to the system (Singh, 1996).

In a typical RCC silo system, of 20,000 tons capacity, grains arriving in trucks / tractor trolleys are mechanically unloaded into the dump pits. This grain is received on an underground belt conveyor which takes it to the inlet of the bucket elevator, which elevates it to the top of the head house where the grain is fed into the large cleaner machines. The grain is thoroughly cleaned and dried before storing. This grain then falls on to a belt conveyor, which can feed all the bins with the help of movable trippers. The quality of the stored grains is monitored regularly by comparing temperature profiles within the grain bulk and is subjected to aeration/ fumigation as and when needed.

Grain is unloaded from the bottom hoppers through control valves. It falls by gravity on to a belt conveyor beneath the valves, which takes it to the feed inlet of the bucket elevator. This elevator takes the grain to the automatic weighing and bagging system where bags are filed with the required weight, machine stitched and automatically loaded into the waiting train/truck for onward movement.

In India, silos have been built with heights varying from 30 to 50 meters and diameters from 6 to 10 meters. Concrete silos with individual bin capacity of 2000 to 4000 tons are in use in different parts of India (Singh and Deol, 1986).

A GIC silo system comprises of silos as containers, silo accessories, and ancillary equipment. The GIC silos can be of flat bottom or hopper bottom design. Hopper bottoms are preferred where turn out is more frequent. If the bins are to be filled once or twice a year, the flat bottom system is more economical and is preferred. The foundation details should be as per site requirements.

Galvanized iron (GI) sheets of appropriate thickness are corrugated in special machines to get GIC sheets of required strength. Holes are punched and sheets are bent using special tools, and machines to predetermined arch-radius corresponding to the bin diameter. Bin diameter should be so selected that when bolted end to end a whole number of bent GIC sheets form a circular ring.

The height of the cylindrical portion of the silo should be so designed that a number of such rings when bolted one on top of another give the designed cylindrical height, and no sheets need to be cut and wasted. The conical roof (30°-35° with horizontal) of the silo is made of GI sheets. Special sealant materials are used for joining the overlap sheets as well as all other bolted joints with structural steel. Special stiffeners are provided to take care of hoop stresses and buckling failures at appropriate locations.

A specially designed manhole cum air vent is also used as inspection window. In flat bottom bins, a suitable door is provided, in the cylindrical side, at a convenient location. Flat bottom silos have a well at the centre and a sweep auger to carry grains from the periphery to the central well for discharging by means of an unloading auger. In hopper bottom silos discharge by gravity is regulated by means of a rotary valve or a discharge gate arrangement. In flat as well as hopper bottom silos a suitably designed blower and perforated ducting system is provided for effective aeration of the product.

A grain spreader can be provided at the top to ensure uniform distribution of grains inside the silo. Grain level indicators are also provided at suitable locations. Grain quality, is monitored by measuring grain temperature profiles, using temperature sensing devices such as thermistors, thermocouples, etc coupled with a temperature recorder.

Community Owned Rural Storage System

In view of the growing foodgrain production and resultant increase in demand for scientific storage, it is important that community storage system for a group of households, or a village or a cluster of villages needs to be encouraged. These community level structures (warehouse or silo) may or may not be individually owned. In the later case the farmer may have the option to either sell the grain to the cooperative at the MSP announced by the government or merely keep it there till he plans to sell it to the cooperative or take it out to sell it elsewhere. In the latter case he will only pay the storage charges. These cooperatives or private grain handlers can handle and/or process grain in bulk, and transport it to the mandi or processor or bulk consumer for further processing/distribution, or even directly export to some other country. This will eliminate the role of middlemen and the resultant savings can be passed on to the cooperative as service charges.

A bulk consumer or processor such as a trader, or a Modern Flour Mill owner, or a community centre, can also benefit a lot by constructing one or more steel silos. The silos can be of flat/hopper bottom type, to optimize the storage and handling charges while reducing the grain losses and minimizing the labour problems. One such system of 420 tons (wheat) capacity has been designed and got constructed by the author, at village Sri Bhaini Sahib (25 km from Ludhiana on Ludhiana-Chandigarh road). The system comprises of three flat-bottom GI silos each of 140 tons capacity, with a 44' long (6" diameter) portable auger conveyor for loading and three horizontal augers (18' long and 5" diameter) for unloading purposes. The system has a handling capacity of about 10-12 tons per hour and is provided with a grain pre-cleaner, and aeration facilities. Processing facilities such as final cleaning of grains; grinding of flour, kneading & rolling of flour dough; making and baking of chapatis is also planned. Similar steps may also be taken by other community level storage, handling and processing centers.

Modernization of Storage System

Population of India is likely to be around 1360 million and foodgrain production to be in the range of 280 to 300 million tons by the year 2020 (Kalam and Rajan, 1998). According to some projections, India can become exporter of foodgrains to the tune of 15 million tons by the year 2020. The major obstacle, to increase in the exports, is the poor quality coupled with high cost of foodgrains as a result of outdated storage and handling techniques. Going by the present trend, the volume of foodgrain procured, stored, moved and distributed by FCI may go up by 50 % over the existing level by 2020. Therefore there is an urgent need to modernize existing storage facilities and to create similar additional storage facilities to take care of additional production (Singh 2000). For practical reasons, this may be phased out over the next 10-15 years.

CONCLUSIONS

1. The farmers should know how to effectively clean dry and store grains, which are meant for their own consumption.
2. The farmers can also clean and dry the grains, meant for immediate or deferred sale, if they are duly motivated and compensated for the extra effort. At present the farmer is charged for two cleanings in the mandi, even if he brings clean grain. This encourages him to not to clean these grains at all.
3. If the farmers are assured of adequate price for good and premium quality grains in the markets, they will do their best to bring high quality clean and dry grains in the mandis, thereby reducing the prevailing conditions of glut and confusion.
4. The Department of Food has to revise its procurement policy by offering higher price for the good and premium quality grains, which may be earmarked as export quality, to fetch higher price
5. The farmers need to be educated about the overall benefits of modern bulk storage and bulk handling system over the outdated bag storage and bag handling system. This will reduce wastage of grains, improve grain quality, improve system efficiency, reduce grain losses, reduce system cost, and promote export potential.
6. The farmer should be encouraged to opt for deferred sale of his produce as raw grain or preferably as a value added product.
7. The government must provide all possible technical as well as marketing knowledge to encourage farmers to become processors and sell value added products to become businessmen in addition to producers of high quality food grains.
8. The government should provide the necessary financial support in the form of subsidies, loans and insurance covers, at least in the initial years, to the farmers or farmer cooperatives, or professional grain handlers and exporters,

or the qualified Agricultural/Engineering/Food graduates who opt for such an enterprise.

9. To create competitive marketing, the government as well as private 'grain handlers and exporters' should encourage contract farming to boost their business and win confidence of the farming community.
10. The government on its part has to create a network of necessary infrastructure facilities for bulk storage, bulk handling and bulk transportation of grains throughout the country and integrate it with the bulk stores throughout the country.
11. The government must come out with a time bound program to gradually replace the bag handling system with cost effective bulk storage handling and transportation system to meet the international quality standards.
12. The grain can be handled and transported in bulk, from mandi to various storage points and the mills. From here, the flour or rice can then be filled into 5,10, 25 or 50 kg bags of convenience, which are machine stitched and transported to final consumer destinations.

REFERENCES

Abdul Kalam A.P.J. and Y.S. Rajan (1998). India 2020. A Vision for the New Millennium. Penguin Books India (P) Ltd., N. Delhi

Anonymous (1998). Report of the Steering Committee on Introduction of Modern Technology in Handling, Storage and Transportation of Foodgrains

Anonymous (2002 A). Markfed in Competition for bulk grain handling. The Tribune, June 11, 2002.

Anonymous (2002 B). FCI Plans 3 storage points. The Tribune, April 23, 2002

Anonymous (2003). Multi-Crop Multi-Year Contract Farming to diversify Punjab's agriculture, The Tribune, Feb. 18, 2003

Mahak G. S. (2003). Johl Calls for Direct export of Paddy, Wheat - Recommends use of Grain Handlers, fixing Minimum Export Price. Hindustan Times, Dec 03, 2003, page 6.

Maseeh M. (1990): Public Warehousing in India. Regional Workshop on, Warehouse Management of Stored Grains, organized by Department of Food GOI, FAO, and UNDP from 19th March to 6th April at New Delhi PP 115-124.

Pingale S.V. (1976). Handling and Storage of Foodgrains. ICAR Publication.

Sawant S.D. (1989). Bulk Storage and Handling of Foodgrains in Bulk Form for Large Units. Proceedings of the National Symposium on 'Handling, Marketing, Transportation and Storage of Foodgrains in India' Deptt. of Processing & Agril Structures, PAU, Ludhiana. pp 144-154.

Singh Rajinder (2000). Relevance of Mechanization of Marketing and Storage Operations in Tune with Hi-Tech Agriculture. Paper presented in 14th National Convention of Agricultural Engineers, organized by The Institution of Engineers (India), Karnataka State Center, Bangalore, on Nov 24-25, 2000.

Singh Y. and G. S. Deol, 1986. Foodgrain Storage for Large Scale Handling. Proceedings of Workshop on Handling, Marketing, Transportation and Storage of Wheat and Paddy in Punjab, held at the Department of Processing and Agricultural Structures, PAU Ludhiana on April 10. pp 32-36.

Singh, Rajinder, 1996. Mechanization of Handling Operations in Wholesale Grain Markets in North India: A Case Study in Punjab. AMA 27(3)57- 61, 64

18

Post-Harvest Management of High Moisture Paddy in Punjab

V.K. Sehgal, V.K. Thapar, M.S. Alam, M. Arora, D.K. Sharma and P.C. Grover

Abstract

High moisture paddy can be dried in a trolley drier at farm level/grain market to reduce losses during various post harvest operations. During last two paddy harvesting seasons quality analysis of eight, most popular varieties of fresh paddy arrivals from about 40 grain markets, scattered all over Punjab, was performed. Monitoring of paddy in grain markets showed that maximum number of samples (62.7%) was in the range of 12-18%, moisture content (w.b.). Over-dried paddy samples (<12% m.c., w.b.) were only 9.1%. The samples with more than 18% m.c ,w.b. (high moisture paddy) were 28.2%, which have to be dried to safe moisture levels. Milling characteristics of the analyzed samples indicated that the broken percentage was highest (25.5%) in PR-113 followed by PR-111 (20.6%) and minimum in Pusa-44 (12%). The highest insect- pest infestation (Sitotroga cerealella (olivier) was observed in PR-108, followed by PR-116, PR-113, PR-111, Lal Mundi, Govinda, PR-114 and Sharbati. The Mycoflora isolated and identified were Aspergillus, Penicillium, Mucor, Rhizopus, Alternaria, Curvularia, Helminthosporium and Cladosporium spp. Maximum fungal infestation were associated with samples of PR-108 variety. To avoid further proliferation of mycoflora and inset pests, paddy should be dried to safe moisture levels for further storage and processing. For this, a trolley drier has been developed by converting a conventional farmer's tractor trolley, into a portable drier, which can reduce 5% moisture of 1 tonne batch of paddy in about 100 minutes, consuming 56 kWh of electrical energy.

INTRODUCTION

High moisture paddy can be dried in a trolley drier at farm level/grain market to reduce losses during various post harvest operations. It has been observed during the last decade in Punjab that the farmers harvested the paddy mechanically at high moisture content to avoid the shattering losses of the paddy in the fields. However, this high moisture paddy during storage may results in the deterioration in the quality of rice after milling in addition to the decrease in the total yield.

MATERIALS AND METHODS

During the paddy-harvesting season of 2010 and 2011, forty-three and thirty seven grain markets respectively, scattered throughout Punjab were selected for taking paddy samples from these markets. These samples were immediately packed in polyethylene bags and brought in the laboratory and their moisture content; milling parameters; insect infestation and the mycoflora associated were analyzed as per BIS standards, and International Seed Testing Association procedures.

To obtain maximum milling yields and to avoid proliferation of mycoflora and insect pests, paddy should be dried to safe moisture levels for further storage and processing. For this, a tractor trolley available with the farmer has been converted into a dryer. The total size of the screen on which the material is to be placed for drying is 4267.2 x 2133.6 mm where as the perforated screen is of the dimension of 4054 x 2042 mm. The dryer consist of a set of three heaters (8 kW each) for heating the air to any desired temperature. The dryer was tested for paddy this year at the farm of Department of Plant Breeding. The capacity / batch was 10 qtls. and was approximately 225 mm thick when loaded on the trolley drier for drying. The dryer was tested at its full load.

RESULTS AND DISCUSSION

A total of 191 and 111 paddy samples were collected in the harvesting season of 2010 and 2011 from 43 and 37 grain markets of Punjab, respectively. Depending upon the requirements of the procuring agencies, miller's requirements and the storage conditions, the paddy samples were categorized in four different ranges of moisture content (Table 1).

In the first range, paddy with more than 18% m.c (w.b.) has to be dried before procurement by the different agencies. The second range of 14-18 per cent is good for the procurement agencies. The third range is better for the millers and the fourth range is good for storage of paddy in the godowns. The data presented in this table indicate that the maximum numbers of samples (43.9 and 53.1 % for the year 2010 and 2011, respectively) were found in the range of 14-18 % m.c. (w.b.). The moisture range of more than 18 per cent have been observed in 28.2 and 12.6 per cent of the samples for the year 2010 and 2011, respectively. During these years 8.8 and 21.6 per cent of the samples were found to have 12-14 %

m.c. The minimum number of samples 9.1 and 12.6 per cent during the subsequent years were in the range of less than 12 per cent moisture. The low moisture content in these paddy samples may be due to the delay in the procurement by different agencies.

Table 1: Range of moisture content of paddy samples in Punjab

Sr. No.		Year 2010		Year 2011	
	Range %	No. of Samples	% of Samples	No. of Samples	% of Samples
1.	> 18	54	28.2	14	12.6
2.	14-18	84	43.9	59	53.1
3.	12-14	36	18.8	24	21.6
4.	< 12	17	9.1	14	12.6

Milling Characteristics

The data on the milling parameters (variety wise) indicate that the husk content of the varieties is in the range of 20.71 to 23.82 percent (Table 2).

Table 2: Milling parameters of paddy samples collected from different districts of Punjab

Sr. No.	Variety	Milling Characteristics					
		Initial m.c. (%)	Husk content (%)	Total yield (%)	Degree of Polish (%)	Head yield (%)	Brokens (%)
1.	Pusa-44	16.83	20.71	72.73	8.28	62.81	12.00
2.	PR-114	16.54	21.29	70.52	5.33	62.65	14.22
3.	PR-111	13.85	22.66	71.44	8.81	56.77	20.57
4.	PR-108	19.09	21.50	73.28	6.59	64.99	11.50
5.	PR-116	16.17	21.91	73.43	6.98	62.04	15.50
6.	PR-106	15.73	21.67	72.47	7.48	60.07	17.58
7.	PR-113	18.09	20.82	73.42	7.27	54.74	25.50
8.	Sharbati	15.15	23.82	70.35	7.63	57.63	18.12
9.	Misc.	15.18	22.19	72.80	6.42	61.07	16.20
	Average	16.29	21.84	72.27	7.19	60.31	16.79

The maximum husk content was found in the variety Sharbati (23.82 percent) which is reflected in the lowest value of total yield i.e. 70.35 per cent. Similarly, minimum husk content of 20.71 and 20.82 per cent in Pusa-44 and PR-116 are reflected in the total yield of 72.73 and 73.42 per cent respectively. However, the data presented in the table indicates that the head yield were maximum in PR-108 (64.99 %) and in Pusa-44 (62.89%) and minimum in PR-113/54.74 percent. Similarly, the broken percentage was observed to be highest (25.5 per cent in respect of PR-113 followed by PR-111 (20.57 per cent) and minimum in Pusa-44 (12.0 percent). The maximum percentage of bran (8.81 followed by 8.28) was observed in the variety PR-111 and Pusa-44, respectively. The minimum percentage of bran was found in the variety PR-114.

Drying of Paddy

The paddy having average moisture content of nearly 18% (d.b.) was loaded on trolley drier for reducing its moisture content upto 13% (d.b.). The dryer was tested at its full load and was observed that during drying the air velocity through the plenum chamber was 1± 0.2 m/s resulting in flow rate of 496.8 m^3/min. During drying the temperature of drying air varied from 27.3°C (ambient temperature) to 54°C (maximum temperature attained), whereas the dryer attained the temperature of 50°C within 30 minutes. The paddy was stirred after every 30 minutes interval for uniform drying. During drying the ambient air temperature varied from 27.3°C to 29°C whereas the humidity of air varied from 35% to 40%. The reduction in moisture content with time was measured by taking samples from five different locations at different time intervals for assessing the uniformity in drying. The samples were taken from the top as well as from the bottom layer to see the variation in moisture content with time (Fig. 1).

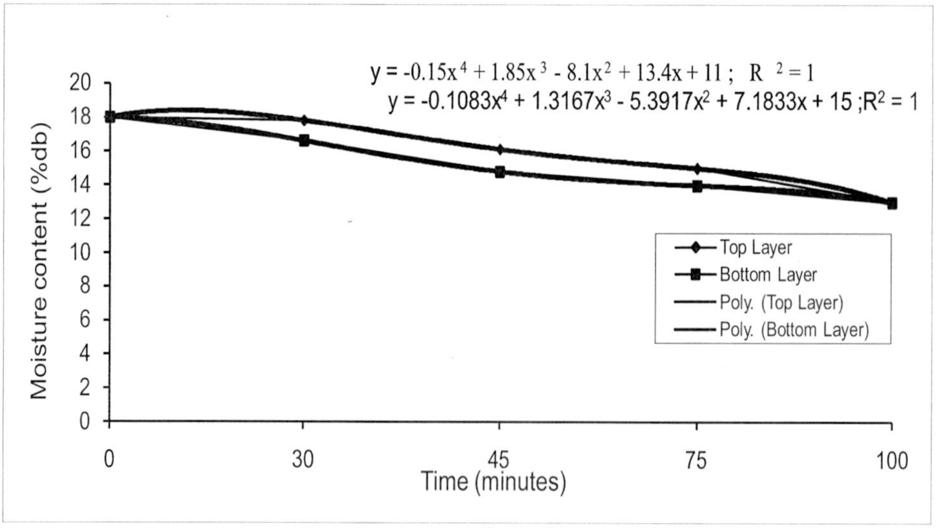

Fig. 1: Drying behaviour of Paddy in a trolley drier

It is very clear from the curve that in the initial stage of drying the moisture reduction is fast in the bottom layer while there is a little decrease in moisture content from the top layer, this may be due to the moisture migration from bottom to the top layer; but after 30 minutes of drying when the dryer temperature reaches approximately around 50°C the moisture removal from both the top as well as bottom layer was at the same rate. The fourth degree Polynomial equation was found to be best fit for the drying behaviour of both the top as well as bottom layer of paddy. The total time taken to reduce moisture content upto 13% (d.b.) was found to be 100 minutes while the total electrical power consumed in reducing moisture content by 5 per cent i.e. from 18%-13% (d.b.) was approximately 56 kWh. Thus the high moisture paddy can be safely dried in a trolley drier with an acceptable quantity and quality.

ACKNOWLEDGEMENTS

The Authors are thankful to Indian Council of Agricultural Research (ICAR) for funding the project.

19

Improvement in Sunflower (*Helianthus annus* L.) Seed Quality in Marginal Lots by Specific Gravity Separator

Satish Kumar, Sadhna Arora, V. K. Sehgal and Sanjeev Uppal

Abstract

Sunflower seeds of variety PSFH- 67 were obtained from the PAU seed farm. These seeds were cleaned with commonly used seed processing equipment. Samples were divided into 2 parts. One part was kept as such for taking the observations of feed (control) and the remaining part was passed through specific gravity separator for its refinement. The various fractions obtained were recovery (heavier, healthy seeds), middlings (intermediate, under size) and rejections (lightweight seeds, wevil eaten, foreign matter). The respective percentages of each fraction were computed. These graded fractions as well as the feed were packed in 2-kg cloth bags and stored under ambient conditions. Different quality parameters viz. moisture content, physical purity, percent germination, thousand grain weight and vigour index were recorded at 30 days interval from Jan, 2010 to Apr. 2010. The physical purity, moisture content, thousand grain weight, percent germination, root length, shoot length and vigour index were found to be higher in recovery and minimum for rejections while moisture content was highest for rejection compared to all other fractions. The physical purity remained more or less constant throughout the period of storage. The thousand-grain weight showed a rapid decrease during first month thereafter it remained constant. There was continuous decrease in percent germination with the period of storage irrespective of the treatment. The recovery showed the minimum decrease in vigour index with the period of storage. Specific gravity separator resulted in increase in purity by 1% while thousand-grain weight increased 1.7-5g/1000 grains. Germination and vigour index increased by 6-12 percent and by 199-295 point, respectively. The grading resulted in improvement in overall quality of sunflower seed.

INTRODUCTION

India is the 2nd largest producer in Asia after China with regard to area and production of sunflower. The sunflower has traditionally been cultivated in Karnataka, Maharastra and Andhra Pradesh. In recent years, its cultivation was taken up in non-traditional states like Punjab, Haryana, UP, MP, Rajasthan, Orissa and TamilNadu. The area under cultivation, production and average yield have been increasing since last 15 years which is evident from the fact that area under cultivation has increased from 1.34 million hectars to 5.98 million hectars (Uppal, 2002). In northern region in contrast, the trend has been reversed. A part of sunflower seed was damaged in godowns by untimely rains in Punjab and Haryana in 1997-98 leading to its disposal by the farmers at throw away prices. In 1998-99, rabi season, the seed did not germinate well in whole northern region with the consequence, the area under sunflower came down by about 10-15 percent. The farmers started losing interest in sunflower cultivation. Under these conditions, the necessity for maintaining quality of sunflower seed was felt.

The cleaning and grading of seed is commonly done by double screen cleaner-cum-grader coupled with indented cylinder separator. However, these are not efficient enough to separate infested grains or heavy imputations like stone which look like grains i.e., having identical shape and size. The specific gravity separator suits to these requirements as the separation is on the basis of unit weight (density) which is higher for sound grains compared to infested grains. Sahnay and Dakshinamurthy (1955) found unsuitability of common vibratory type air screen cleaner in separating infested grains from bulk wheat grain sample. An entoleter was able to separate upto 95% of infested grains. Flower and Lim (1959) observed maximum effectiveness of separator at an incline of 15°C and it decreased thereafter. Uichanco and Lantin (1972) reported that purity was affected more by density than by output in a rotary screen separator. Patil and Sarode (1988) found that heavy and medium weight fractions of wheat seed from specific gravity separator were superior in percent germination and vigour index. Kaushal et al (1991) observed that percent seed recovery decreased with increase in sieve size whereas percent seed germination and physical purity followed reverse trend. Prasad et al (1994) observed that sunflower seeds with specific gravity greater than 0.6 were superior in most quality attributes viz. Seed vigour, percent germination, germination rate and electrical conductivity. Keeping in view the above facts, the present investigation was undertaken to study the improvement in sunflower seed quality by specific gravity separator.

MATERIALS AND METHODS

Sunflower of variety PSFH- 67 seeds were obtained from the PAU seed farm. These seeds were cleaned with usual seed processing equipment. Samples were divided into 2 parts. One part was kept as such for taking the observations of feed (control) and the remaining part was passed through specific gravity separator for its refinement. The various fractions obtained were recovery (heavier, healthy seeds), middlings (intermediate, under size) and rejections (lightweight seeds, weevil eaten, and foreign matter). The respective percentages of each fraction were computed. These graded fractions as well as the feed were packed in 2-kg cloth bags and stored under ambient conditions. Different quality parameters viz. moisture content, physical purity, percent germination, thousand grain weight and vigour index were recorded for all the treatments at 30 days interval. Two replications from each treatment were taken. Moisture content was determined by standard oven method (Chakravarty, 1995). Physical purity in percent was calculated by taking 120 gm samples and separating the inert matter, damaged seeds and pure seeds from the samples. Thousand seeds were counted on electronic balance separator and the average thousand-grain weight was computed in terms of g/1000 grains. For calculating percent germination, 100 seeds were taken and placed in between moist germination papers with sufficient space between them. These were kept at a temperature of 25°C in an incubator. Samples were taken out after 7 days interval and the number of germinated seeds were counted and germination percent was computed. From the above-mentioned sample, 15 germinated seeds were taken and their average root length and shoot length was measured in centimetres. Vigour index was calculated using the formula (Aggarwal, 1980).

Vigour index = [Average root length (cm) + Average shoot length (cm)] X Germination percent

Description of Specific Gravity Separator

Specific gravity separator consists of a triangular shaped perforated table having reciprocating motion. The table is tilted towards the front and towards the left such that a sphere placed on it would roll in the direction of gravity flow. The material is introduced from feed box. The air is blown through the perforated floor at such a rate that the material is partially lifted from the contact with the floor. The lighter & smaller particles are lifted somewhat higher and float down the table towards the discharge end. The heavier and longer particles are not lifted by the air. The oscillating motion moves them in the direction of conveyance and is discharged at the right end of the table. Other materials that are partially lifted are discharged at intermediate points. The steep table pitch can permit high capacity. It is especially useful for removing lightweight infertile, infested seed from seed stock, which cannot be separated by other devices.

RESULTS AND DISCUSSION

Machine Parameters

The deck area was found to be 0.1368 m^2. The airflow rate was fixed at 350 m^3/hr for sunflower. The end-to-end slope and side-to-side slope were 6:1 and 19:1, respectively. Sunflower seed was fed at the rate of 20 kg/hr. All these machine parameters were chosen to have the best performance from the specific gravity separator.

Grade fraction

The different fractions obtained from specific gravity separator were found to be approximately 66 percent, 28 percent and 6 percent for recovery, middlings and rejections respectively. This implied that it is the middlings, which need to be further brought down so as to improve the recovery.

Moisture content

It was found that the moisture content for rejections was highest in the beginning (8.5%) whereas in remaining treatments it was more or less same (7.9%). There was same quantum of decrease in moisture content in all the treatments. The decrease in moisture contents was due to drier environmental conditions.

Thousand Grain Weight

There is continuous decrease in 1000-grain weight with the period of storage irrespective of the treatment. This is verified by decrease in 1000-grain weight. Drier conditions during the period of storage resulted the grains to give off some moisture to the environment. There was a rapid decrease in 1000-grain weight during first month of storage thereafter the weight remained more or less constant till the end of storage due to colder environment. The decrease during first month of storage was 4.7 and 8.22 – 9.12% in feed and graded fraction, respectively. The feed showed the lowest decrease in grain weight. The thousand-grain weight of recovery was maximum while it was minimum in rejections as it contain broken and weevil eaten seeds.

Physical Purity (%)

As seen from Fig. 1 it is clear that physical purity did not have much effect in all treatments during the entire period of storage. The physical purity was highest in recovery followed by feed and middlings and least in rejections. This is because during grading with specific gravity separator the inert matter and light material had moved towards the rejection outlet; as a result its physical purity was maximum for recovery and minimum for rejection. The physical purity remained around 98% in all except rejections, which had physical purity less than 90% after three months of storage.

Percent Germination

The average values of percent germination in all the treatments are presented in Fig. 1. It was observed that the germination was maximum for recovery and minimum for rejection among all the treatments during the entire period of storage. This implied that the germination percentage decreased as we move away from pure seed (recovery) towards rejections. This is due to the fact that the seeds in the recovery were sound whereas rejections contained broken and weevil eaten/ infested seeds. The germination was also higher in recovery compared to that of feed because of proper grading and healthy seeds. However, the germination percentage tend to decrease with the period of storage in all treatments with minimum decrease (27.24%) observed in the feed. Among the graded fractions, minimum decrease of (31.1%) was observed in recovery whereas the decrease was 37.5 percent and 54.05 percent in middlings and rejections respectively. Since in oil seeds the germination decrease is generally high during extended period of storage hence it verified our findings (Justice and Bass, 1978).

Shoot Length and Root Length

The shoot length and root length were observed to be highest for recovery and minimum in rejections. The feed had intermediate values at all duration of storage because it had a mixed lot. It was further observed that shoot length and root length decreased with time. The rate of decrease in shoot length and root length had an erratic trend in all the treatments. The rejections have maximum decrease in shoot length/root length (18.68/18.58%) while the middlings accounted for minimum decrease in root length/shoot length (9.22/2.21%). The feed and recovery had intermediate values.

Vigour Index

The average values of vigour index in different treatments are presented in Fig. 2. It was observed that vigour index was highest in recovery and minimum in rejection among all the treatments during entire period of storage. This is because vigour index is directly affected by germination, root length and shoot length, which were higher in recovery and lower in rejections.

The feed had intermediate values at all durations of storage because it had a mixed lot. The vigour index decreased with the period of storage in all the treatments. The samples of recovery and feed were reported to have minimum decrease (35%) among all the treatments. As we move away from pure seed towards impure seeds the vigour index decrease was 41.2 and 62.6% in respective treatments. In feed and recovery the rate of vigour index decrease was 11.6 percent per month whereas it was 13.93 and 20.8% per month in middlings and rejections. Thus it was concluded that grading results in increase in vigour index and there is further need to improve the efficacy of grader to raise the vigour index.

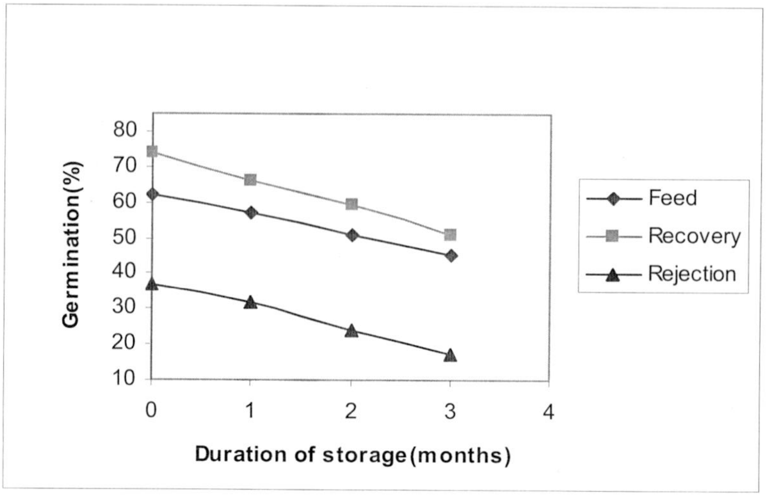

Fig. 1: Percent Germination in different seed fractions of Sunflower

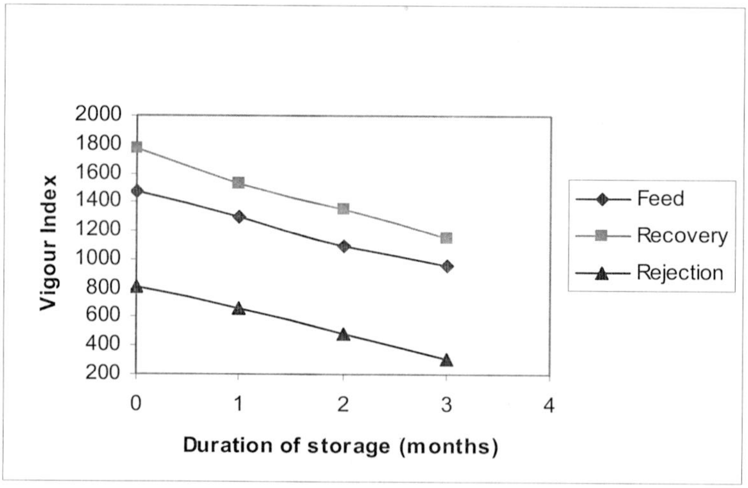

Fig. 2: Vigour index in different seed fractions of sunflower

As shown in Table 1, the physical purity, moisture content, thousand grain weight, percent germination, root length, shoot length and vigour index were found to be higher in recovery and minimum for rejections while moisture content was highest for rejection compared to all other fractions. The physical purity remained more or less constant throughout the period of storage. The thousand-grain weight showed a rapid decrease only during first month thereafter it remained constant. There was continuous decrease in percent germination with the period of storage irrespective of the treatment. The recovery showed the minimum decrease in vigour index with the period of storage. Purity results in an increase by 1% while thousand-grain weight increased 1.7-5g/1000 grains. Germination and vigour index increased by 6-12% and by 199-295 point, respectively. The grading resulted in improvement in overall quality of sunflower seed.

Table 1: Summary of results

Quality attribute	Comparative improvement
Moisture content (%)	No change (0.2)
Physical purity (%)	1
Thousand grain weight	1.7-5
Germination (%)	6-12
Root length (cm)	0.01
Shoot length (cm)	0.21
Vigour index (%-cm)	199-295

REFERENCES

Aggarwal RL (1980) Seed technology. Oxford & IBH Publishing Co. Pvt. Ltd., New Delhi.

Chakraverty, A (1995). Postharvest technology of cereals, pulses and oil seeds (3rd edn.). pp: 26-27, Oxford & IBH Publishing Co. Pvt. Ltd. New Delhi

Flower BT and Lim SC (1959). The influence of various factors on effectiveness of separation of finely divided solids by a vibrating screen. Chemical Engg. Sciences, 10:163-170.

Justice OL and Bass LN (1978) Principles and practices of seed storage. US Govt. Printing Office, Washington DC.

Kaushal R T, Zade V R, Chauke R P and Changade S P (1991) Influence of Seed Size on Seed Quality of Sunflower. Ann. of Pl. Physiol, 5(2):288-290.

Modi B S and Parker R E (1979) Performance of a Paddy Separator. J. of Agric. Engg. 4:183-190.

Patil VN and Sarode SD (1988) Seed Quality Studies in Wheat as Influenced by Specific Gravity Separator. Seed Research 16:(1):114-116.

Prasad SR, Reddy A S and Seenappa K (1994) Influence of Specific Gravity on Vigour and Viability of Sunflower seeds. J. Oil Seeds Research, 11(2):196-200.

Sahay KM and Dadshinamurthy A (1995) Studies on Separation of Infested Grains from a bulk wheat grain. J Fd Sci and Tech., 32(3):252-258.

Uichanco VA and Lantin RM (1972). Determination of Optimum Design Parameters of Rotary Screen Cleaners using Dimensional Analysis. The Phillippine Agriculturist. 56(9 & 10): 219-308.

SECTION- V

PRODUCT DEVELOPMENT: STATUS AND NEW STRATEGIES

20

Plant Breeding — An Orientation Towards End Product Quality

A.K. Bakshi and G.S. Mahal

Abstract

The establishment of All-India Coordinated Wheat Improvement Project in 1965 coincided with the introduction of high yielding semi-dwarf wheat varieties. Since then India has made very impressive progress in increasing wheat production over the past three and half decades. India now ranks as the second largest wheat-producing nation in the world. In fact from a state of "Ship to mouth" in 30 years or so India has become a nation of surpluses. The significant contribution made by the project in helping the increased wheat production cannot be ignored. Through coordinated research and development, nearly 200 wheat varieties have been developed during the last 35 years in India. However, only a small number of them produced without backing of specific quality breeding programs excel in quality traits also. The increasing demand of baked and pasta products in the country along with economic liberalization and global trade have offered opportunities for better utilization of the surplus wheat. Thus, in the current decade and beyond, wheat quality needs uppermost attention to meet the trade requirements for domestic and export purposes. Therefore, there is a need to reorient the wheat improvement programs in the country.

INTRODUCTION

The establishment of All-India Coordinated Wheat Improvement Project in 1965 coincided with the introduction of high yielding semi-dwarf wheat varieties. Since then India has made very impressive progress in increasing wheat production over the past three and half decades. Wheat output in India during this period has witnessed almost six time increase from around 12.2 m tons in 64-65 to 70.3 mt in 2002-03. During the same period area under wheat has almost doubled from 13.4 mha to 24.2 mha while productivity has shown threefold increase from 913 kg/ha to 2900 kg/ha. India now ranks as the second largest wheat-producing nation in the world after China and contributes approximately 12% to the world wheat basket from an area of 11% under wheat in the world. In fact from a state of "Ship to mouth" in 30 years or so India has become a nation of surpluses. The significant contribution made by the project in helping in increased wheat production cannot be ignored.

To keep pace with the rapidly growing population which will be 1.25 billion and to meet the projected demand of wheat for domestic consumption likely to be 109 mt by the year 2020, in the wheat improvement programme emphasis has always been to develop high yielding genotypes that are responsive to high management while keeping the strong base in disease resistance. Through coordinated research and development, nearly 200 wheat varieties have been developed during the last 35 years in India. However, only a small number of them produced without backing of specific quality breeding programmes excel in quality traits also. The increasing demand of baked and pasta products in the country along with economic liberalization and global trade have offered opportunities for better utilization of the surplus wheat. Now there is a need to develop product specific varieties combining high yield potential and disease resistance as well. Thus, in the current decade and beyond, wheat quality needs uppermost attention to meet the trade requirements for domestic and export purposes. Therefore, there is a need to reorient the wheat improvement programme in the country.

More than 90% of the wheat in India is consumed in the form of chapattis. The tall varieties grown before the introduction of semi-dwarfs were, therefore, invariably possessed good chapatti making properties. On the contrary initial introductions from Mexico were dark red colored and hence were acceptable to chapatti eating public with reluctance. Concerted efforts jointly put forth by breeders and quality workers successfully transformed the Mexican red grain into the amber golden colour, a choice of millions of chapatti consumers. Most of the varieties selected and released/ identified since 1965 are comparable for chapatti making to the best ones available earlier such as C306, C591 and hybrid 65 and better than the first group of semi dwarfs.

Globally wheat is mostly consumed in the form of bread. Among the earliest developed varieties K 68 was the only variety suitable for bread making. Indian wheat's fetched a lower price in the export market as it was generally believed that they were not of a very good quality. Lack of arrangement for carrying out technological tests on milling and baking quality of cereals in earlier years was mainly responsible for it. With the setting up of a few cereal research laboratories fully equipped with modern milling, baking and cereal testing equipment has enabled to identify some promising varieties for bread quality. Some of them are suitable for chapatti making also.

The chapatti making quality of wheat, which is ruled by the domestic market, may have to accommodate other types of wheat that are suited for product development. It will be necessary to develop wheat genotypes that are better suited for export and processing. Also the quality of Indian wheat's may have to match that of Australian and North American wheat in bread making. Therefore, information on dough properties is important in relation to specific end uses such as for bread making, biscuit and cakes etc.

The country now has reached the stage when it can export durum wheat, which has the similar potential as that of basmati rice. For meeting the potential markets for Indian durum, it is essential to look into various uses of durum in Middle East and North African countries. Therefore, the strategy for the varietal development has to keep in view the needs of the clients. This calls for fine-tuning of the breeding programme to meet the quality requirements and the demands of the consumer and industry.

The wheat breeding programmes throughout the country have been primarily focusing on yield improvement and disease resistance from the initial breeding stage. The quality aspect, however, is deferred to a stage when the genotypes attain uniformity. Very few wheat breeding centres in the country are equipped to carry out such tests. Setting up of quality laboratories, having modern facilities like Electrophoresis, Gel-scanners, Alveographs, Kjeldhal Auto system, Mixograph, Texture Analyzer, Mills, etc. at all the major centres is essential for rapid screening and categorization of various end uses. Kaul (1969) outlined a scheme for handling breeding materials primarily for the evaluation of bread making.

Keeping in view the major use of wheat in India for chapatti making there is also a need to carry out wheat improvement work in which chapatti making aspects form an essential selection criterion (Gill, 1979).

The All India coordinated wheat improvement project conducts organized multi location yield trials for various growing conditions in six agro-ecologically different zones throughout the country. As per the existing procedure, materials are entered in the National Initial Varietals Trials (NIVT's) on quota basis by breeders working at different research centres on the basis of yield performance and disease resistance without any quality consideration. Superior genotypes identified following multi location testing are promoted to Advanced Varietals Trial (AVT). Better performing genotypes are retained in the AVT.

The promotion/selection criterion is based mainly on the yielding ability and rust resistance of the genotypes. Though data on quality parameters is also available, it is not considered while promoting/retaining genotypes. For developing genotypes suitable for export and processing industry the existing promotion/retention criterion needs modification. The desirable levels of different quality parameters should be an integral part of this criterion along with yielding ability and rust resistance. Few years back a quality trial comprising slightly low yielding but qualitatively desirable genotypes also used to be conducted. This trial was subsequently discontinued. To avoid loss of qualitatively desirable materials such a trial needs revival.

The surplus wheat scenario has opened up new opportunities for the wheat based food industries in North-western India. Permitting of export of wheat will also promote the better economic return to the farmers and thereby encourage them to produce more grain at a competitive rate. In India focused efforts have never been made to develop wheat varieties with hard texture. Hardness of grains is an important criterion for millers since such grains yield greater amount of flour. Water absorption capacity of the flour increases appreciably if the grain is hard and this in turn, increases shelf life of bread and makes it more pliable. Since the hectolitre weight value is related with flour recovery, genotypes with high hectolitre weight should be developed through organized efforts at various breeding centres.

From the ensuing discussion it is very clear that wheat quality needs uppermost attention in times to come to meet domestic and export market requirements. The increase in domestic demand of baked and pasta products, economic liberalization and global trade have offered opportunities for better utilization of wheat surplus. Wheat varieties with different protein levels and hardness are required to fit into diversified bakery products. In North America and Europe different categories of wheat are grown by the farmers and traded as per quality. In Canada alone there are 7 classes of wheat that fit into the needs for bread, biscuits, pasta, cakes, pizza etc. However, in India the wheat is procured and sold as FAQ wheat. To keep pace with the global scenario India would have to expand its horizon and vision to match the International standards for wheat.

With this in mind the Indian quality breeding programmes need reorientation and strengthening. For most of the quality parameters the genetic base is very narrow. There is need to generate variability for quality characters by resorting to hybridisation with exotic sources and introgression from related wild progenitor species. Until now the quality testing had been operational at the advanced stages of breeding, where only the quality of variety is assessed but cannot be fixed as desired. To develop varieties that suit chapatti, noodle, biscuit or bread quality, the rigorous screening shall be needed in the early generation segregating materials.

Heterogeneity is highest in the early generations and by performing half kernel analysis one can find out the materials that suit a particular purpose. For attaining this some centralized laboratories furnished with latest instruments and analytical techniques are required so that the varieties that have high yield, quality and stability could be found out. Atta and chapattis being the national priority agenda, work at molecular level is needed to develop materials that shall produce chapattis like the desi wheat's. Also there is a need to patent varieties and technologies used for atta production, otherwise the MNC's shall obtain the patents and debar the Indian industry from using the old traditional practices for production and sale of atta. The operational programme at All India Coordinated level needs reorientation to avoid losses of valuable quality materials.

REFERENCES

Austin, A. and Ram. A. (1971). Studies on chapatti making quality of wheat. ICAR Tech. Bull. (Agric.) No.31.

Gill, K.S. (1979). Research on wheat and triticale in Punjab. PAU Publication.

Gill, K.S. (1979). Research on dwarf wheat's. ICAR Publication.

Kaul, A.K. (1969). Tests for quality in early generation of wheat. Indian J. Genet. Pl. Breed., 29:348-52.

Murty, G.S. (1958). Wheat Varieties. ICAR, Review Series No. 13.

Nagarajan, S., Verma R.P.S. and Tyagi, B.S. (1997). Vision-2020. ICAR Publication.

Tandon, J.P. and Sethi A.P. (1986). Twenty five tears of coordinated wheat research. 1961-1986. ICAR Publication.

21

Physico-Chemical, Milling and Rheological Characteristics of Indian Wheat and Triticales for Pretzel Making

H.R. Naik, K.S. Sekhon, H.P.S. Nagi and G.M. Beigh

Abstract

Four bread wheat varieties (PBW-138, PBW-299, PBW-343 and PBW-373), two of durum wheat (PDW-215 and PDW-233) and two of triticale (TL-419 and TL-1210) were milled to obtain 70 percent extraction flour. The grains and the flour samples of all the varieties were evaluated for proximate composition, rheological properties and pretzel making. Physical properties of durum wheat were better followed by bread wheat and triticale. Durum wheat were harder, had higher grain weight than bread wheat and triticale. Protein, ash, fat, total sugars and enzyme activity were higher in triticale flour than bread and durum wheat flours while durum flour had higher pigments (ppm, b-carotene). Bread wheat flour was intermediate between durum and triticale flours with respect to fineness. Rheological studies revealed triticale flour to be the weakest flour. Pretzels prepared by using optimized recipe and method revealed PBW-343 and PBW-373 best for pretzel making.

INTRODUCTION

Wheat is the second most important cereal crop of India after Rice with an annual production of 73.5 million tonnes in 2001-2002 (Directorate of Economics and

Statistics, 2002). Among cereal, wheat is known as 'stuff of life' or the king of cereals because of its pre-eminent suitability for preparation of large number of palatable products, such as, bread, biscuits, breakfast cereals and snacks foods. This arises from the fact that wheat contains a unique visco-elastic property, which renders the dough extensible.

Triticum aestivum (Bread wheat) and *Triticum durum* (Macaroni wheat) are two commonly grown wheats. Triticale (*Triticosecale wittmack*) is a man made cereal produced by cross between (*Triticum* spp.) and rye (*Scale cereal* L.). Total estimated area under triticale is reported to be 500 thousand hectares in the world (Varughese *et al.*, 1996). All the three types of wheat have been used for the production of various baked products. At present export of wheat from India is almost exclusively in the form of unprocessed whole grains. Production of value added products is the need of the hour in order to boost the profitable utilization of surplus wheat available (Patnaik, 1992). Due to urbanization, there is a great demand of pre-cooled, and ready-to-serve products, which suit the pockets of different strata of population. Rightly, a major thrust is being made in the development and large-scale production of these types of value added products (Bakshi and Binger, 1991). Pretzel is one of them. Pretzels are a very tasty, low calorie, nutritious cooked and baked snack products made from wheat. Typical ingredients used are flour, water, yeast, malt, shortening, soda and salt. Flour requirements are usually low protein (9-9.5%) with high ash content.

Perusal of literature shows only limited studies elsewhere and no study seems to be available for pretzel making from Indian wheats. Using standard recipe and method developed for pretzel production by (Ramandeep, 1998), the studies were taken with the objectives:

i) To study physico-chemical, rheological and milling characteristics of Indian wheat

ii) To study the effect of variety on pretzel making

MATERIALS AND METHODS

Test Materials

Four bread wheat (PBW-138, PBW-299, PBW-343, PBW-373), two durum wheat (PDW-215, PDW-233) and two triticale (TL-419, TL-1210) varieties were obtained from different experimental farms of Department of Plant Breeding, Punjab Agricultural University, Ludhiana. These varieties were grown under dissimilar agronomic conditions of Punjab. The samples were thoroughly cleaned and stored in metallic bins for subsequent use in investigation.

Milling of wheat

Test samples (2 Kg) were conditioned at 15, 16, and 17 percent moisture for bread, wheat, durum wheat and triticales and rested for 48 hours before milling in the Buhler Pneumatic Laboratory Mill (MLU-202) under appropriate conditions of feed rate, roll settings and siftings. Base flours of 70 percent extraction were obtained from all the varieties by mixing break (B1+B2+B3) and reduction (R1+R2+R3) fractions of Buhler Mill. The flour was stored in labeled air tight (Pet) Jars for use in the analysis.

Physico-chemical characteristics

Samples of grain and flour were analyzed for 1000- kernel weight, hectolitre weight, perling index, moisture content, crude protein, ash, crude fat, pigments, falling number and sugars using standard AACC procedures (1993). Density (g/ml) of grains was determined as described by Bhattacharya *et al* (1972) and Starch by Lane and Eynon method (1923). Particle size distribution of flour was determined by Henry Simon Laboratory Methods (1985).

Rheological studies

Dough characteristics of wheat and triticale varieties were studies by using Barbander Faringograph and Amylograph. Farinogram values were derived from a 15 min mixing curve for 50 g flour (14% moisture content basis) and sufficient distill water to give a maximum dough consistency centered on the 500-BU line according to AACC method (1993). Amylogram values were derived from 70-gram flour (14% moisture content basis) and 450 ml of distilled water. The instrument was set for 1.5°C rise in temperature per minute with an initial adjustment of temperature to 30°C.

Pretzel preparation

Baker's yeast, shortening (Refined cotton seed oil), salt and ammonium bicarbonate were obtained from local market. Malt extract was prepared in the laboratory and the pretzels were prepared as per the recipe and method optimized by Ramandeep (1998).

i) Preparation of malt extract: Malted barley was ground in the hammer mill and the ground material was added to water in 1:1 ratio. Steeping was done at about 40°C for 20 minutes and the material was heated to a temperature of 65°C with constant stirring for 20 minutes. The slurry was then filtered through a muslin cloth. After separating the liquid, two more similar extractions were carried out. All the extractions were combined and kept in a separating funnel for six hour. Supernatants collected were concentrated at mild heat over hot plate to about 65°Brix with constant stirring.

ii) Recipe and method for pretzel preparation:

RECIPE

Flour	100.0 g
Water	50.0 ml
Compressed yeast	0.5 g
Salt	2.0 g
Ammonium bicarbonate	0.4 g
Refined oil	3.0 ml
Malt extract	2.0 g

METHOD

Mixing	7 minutes
↓	
Shaping	Thin, oval
↓	
Proofing	10 minutes
↓	
Cooking	1% NaOH at 88°C for 15 seconds
↓	
Baking	218°C for 6 minutes
↓	
Drying	121°C for 25 minutes

To find out the best wheat variety for pretzel production, the product was organoleptically evaluated using 9-point hedonic scale for appearance, texture, flavour and overall acceptability.

Statistical analysis

The data collected on different characteristics were analysed with the help of CRD (Completely Randomized Design) and least significance difference (5% level) between means were calculated. All the data was expressed at 14% moisture basis unless otherwise mentioned.

RESULTS AND DISCUSSION

Physico–chemical characteristics of grain

It is evident from Table 1 that 1000-kernel weight (43.04, 40.84, 39.14 g), test weight (82.36, 80.97, 70.46 kg) and density (1.86, 1.74, 1.64 g / l) were observed significantly highest in durum wheat followed by bread wheat and triticale respectively. However, values for pearling index were statistically similar for bread (78.11% overs) and durum (77.34% overs) wheat but higher than triticales (64.37% overs).

Higher values of these parameters in durum wheats indicated them plumper, denser and strong wheats and lowest in triticales were to be attributed to some amount of shriveling of grains. The results were in consonance with those reported by Kumar *et al* (1979), Bakhshi and Bains (1987), Dexter *et al.* (1991) and Nemeth *et al.* (1994). Results indicated that bread and durum wheat were harder. Barlow *et al.* (1973) explained grain hardness due to light adhesion between starch granules and storage proteins of hard wheats.

Protein content of triticale (11.25%) was significantly highest followed by durum (10.54%) and bread (9.45%) wheats, respectively (Table 1). The results were in consonance with those reported by Harinder and Bains (1989) and Hira *et al.* (1991) for Indian wheat varieties. However, Sehgal *et al.* (1982) reported 12.5 – 13.2 per cent in triticales. No variation was observed with respect to ash and fat content in different wheats, however, higher ash (2.02%) in triticales indicated higher proportion of bran in them. Singh *et al* (1988 and 1992) reported similar results. When grain weight or test weight becomes low, the bran to endosperm ratio becomes higher leading to enhanced ash values. Starch content of bread, durum and triticales were 59.44, 60.58 and 57.57%, respectively.

Physico–chemical characteristics of 70% extraction flour

Table 2 revealed that protein (10.05, 9.33, 8.78%), ash (1.04, 0 .74, 0.72%), and fat (0.68, 0.67, 0.52%) were higher in triticale followed by bread and durum flours. However, reducing sugars (1.08, 0.99, 0.97% glucose), pigments (7.78, 4.12, 2.41 β- carotene) and falling number (555.67, 462.08, 232 sec) were recorded higher in durum flour followed by bread and triticale flours. Total sugars (5.72, 5.56, 5.56%) and starch (64.15, 62.91, 62.70%) were again found higher in triticales followed by durum and bread wheat flours while non reducing sugars were 4.46, 4.35 and 4.27% for triticale, bread, and durum flours. The findings recorded fall in range reported by Bakshi and Bains (1987), Nemeth *et al.* (1994), Leon *et al.* (1996), Singh *et al* (1983 & 1988), Gill *et al* and Sekhon *et al.* (1980). The values for pigments reported by Bakhshi and Bains (1987) and Kumar *et al.* (1979) are lower than those recorded in the investigation.

Higher amount of ash content in triticale may be due to the reason that Indian triticale are at developing stage and had a lot of shriveled kernels. Falling number results indicated that bread and durum wheat were sound and deficient in alpha amylase than triticales. Bread wheat and triticale flours revealed lower amount of alpha amylase than those reported by Venketeshwara Rao *et al* (1977), Lukow and Mcvetty (1991) and Leon *et al* (1996). However, Falling number value for durum wheat was in consonance with the findings of Boyacioglu (1991).

Table 1: Physico-Chemical Characteristics of bread, durum and triticale wheats

Class/ Variety	1000-kernel weight (g)	Hectolitre weight (kg)	Density (g/ ml)	Pearling index (% overs)	Protein (%)	Ash (%)	Fat (%)	Starch (%)
Bread wheat								
PBW-138	42.10	82.55	1.86	78.50	9.19	1.61	1.17	58.20
PBW-299	43.00	84.36	1.89	79.97	9.09	1.45	1.20	59.11
PBW-343	43.90	82.36	1.88	75.00	10.47	1.61	1.35	62.76
PBW-373	34.48	74.61	1.33	78.96	9.04	1.85	1.29	57.70
Mean	40.84	80.97	1.74	78.11	9.45	1.63	1.25	59.44
Durum wheat								
PDW-215	48.53	83.99	1.86	75.00	10.25	1.57	1.28	58.00
PDW-233	37.65	80.73	1.86	79.67	10.83	1.89	1.05	63.16
Mean	43.04	82.36	1.86	77.34	10.54	1.73	1.17	60.58
Triticale								
TL-419	38.32	71.98	1.79	70.73	11.50	2.20	1.21	56.65
TL-1210	39.95	68.94	1.48	58.00	11.00	1.84	1.18	58.48
Mean	39.14	70.46	1.64	64.37	11.25	2.02	1.20	57.57
CD(0.05)	1.37	0.60	0.04	1.67	0.007	0.09	NS	0.07

Table 2: Physico-chemical composition of bread, durum and triticale flour at 70% extraction rate

Variety	Protein (% N x 5.7)	Ash (%)	Fat (%)	Reducing sugars (% glucose)	Non-reducing sugars (% sucrose)	Total sugars (% glucose)	Starch (%)	Pigments (ppm B-carotene)	Falling number (sec)
Bread wheat									
PBW-138	9.04	0.88	0.52	1.01	3.62	4.82	59.45	3.91	475.67
PBW-299	8.49	0.86	0.70	1.02	4.46	5.71	59.31	2.91	516.33
PBW-343	8.84	0.58	0.44	0.91	4.70	5.86	65.59	4.82	382.00
PBW-373	8.74	0.57	0.43	1.01	4.61	5.86	66.43	4.82	474.33
Mean	**8.78**	**0.72**	**0.52**	**0.99**	**4.35**	**5.56**	**62.70**	**4.12**	**462.08**
Durum wheat									
PDW-215	9.29	0.74	0.54	1.14	4.94	6.34	61.97	5.72	583.00
PDW-233	9.36	0.74	0.79	1.02	3.59	4.81	63.84	9.63	526.33
Mean	**9.33**	**0.74**	**0.67**	**1.08**	**4.27**	**5.56**	**62.91**	**7.68**	**554.67**
Triticale									
TL-419	10.20	1.01	0.70	0.91	4.94	6.11	63.60	2.71	216.00
TL-1210	9.90	1.07	0.66	1.02	3.97	5.33	64.70	2.11	248.00
Mean	10.05	1.04	0.68	0.97	4.46	5.72	64.15	2.41	232.00
CD(0.05)	0.01	0.05	0.05	NS	0.14	0.16	1.98	0.11	27.07

Particle size distribution of flour

Bread, durum and triticale wheat differed significantly with respect to particle size for less than 50 and more than 100 microns (Table 3). The highest aggregate flour particles were observed in the range of 50-100 microns in durum wheats (41%), followed by bread (40%) and triticale (39.5%). Particle size aggregate of less than 50 microns was found to be 38 per cent in triticale followed by 21.8 and 14.0 per cent in bread and durum wheats. However, durum had highest value of aggregate particle size (more than 100 microns) 45 per cent followed by 38.25 and 22.5 per cent in bread and triticale wheats. Results indicated that triticales had more fine flour in the range less than 50 microns where as durum flour composed of larger particles. Bread wheat had intermediate particle size flour and its varieties varied significantly with respect to flour fineness.

Table 3: Effect of variety on particle size distribution of flour at 70% extraction rate

Variety	Less than 50 microns	50-100 microns	More than 100 microns
Bread wheat			
PBW-138	21	39	40
PBW-299	20	39	41
PBW-343	27	33	40
PBW-373	19	49	32
Mean	21.8	40.0	38.25
Durum wheat			
PDW-215	13	41	46
PDW-233	15	41	44
Mean	14.0	41.0	45.0
Triticale			
TL-419	39	37	24
TL-1210	37	42	21
Mean	38.0	39.5	22.5
CD (0.05)	0.61	1.6	1.4

Rheological characteristics

i) Farinographic properties

Durum flour showed highest water absorption (59.55 %) followed by bread (58.44%) and triticale (50%) flours respectively (Table 4 and Fig.1).

Table 4: Effect of variety on Farinographic properties of 70% extraction flour

Variety	Water absorption (%)	Arrival time(min)	DDT (min)	Stability (min)	MTI (BU)	Degree of softening (BU)
Bread wheat						
PBW-138	62.12	1.5	2.20	3.5	70.0	105
PBW-299	57.90	1.4	1.90	1.4	65.0	87
PBW-343	56.66	1.5	2.00	2.5	90.0	110
PBW-373	57.06	1.6	1.97	3.7	60.0	85
Mean	58.44	1.5	2.02	2.8	71.25	96.75
Durum wheat						
PBW-215	59.64	2.0	3.50	16.0	5.0	25.0
PDW-233	60.26	2.1	3.40	23.30	20.0	10.0
Mean	59.55	2.05	3.45	19.65	12.5	17.5
Triticale						
TL-419	51.34	1.2	1.6	1.30	11.5	115
TL-1210	48.66	1.0	1.4	0.60	100.0	155
Mean	50.00	1.1	1.5	0.95	107.5	135.0
CD (0.05)	2.52	0.61	1.01	1.33	11.49	15.34

DDT = Dough Development Time; MTI = Mixing Tolerance index; BU = Brabender Units

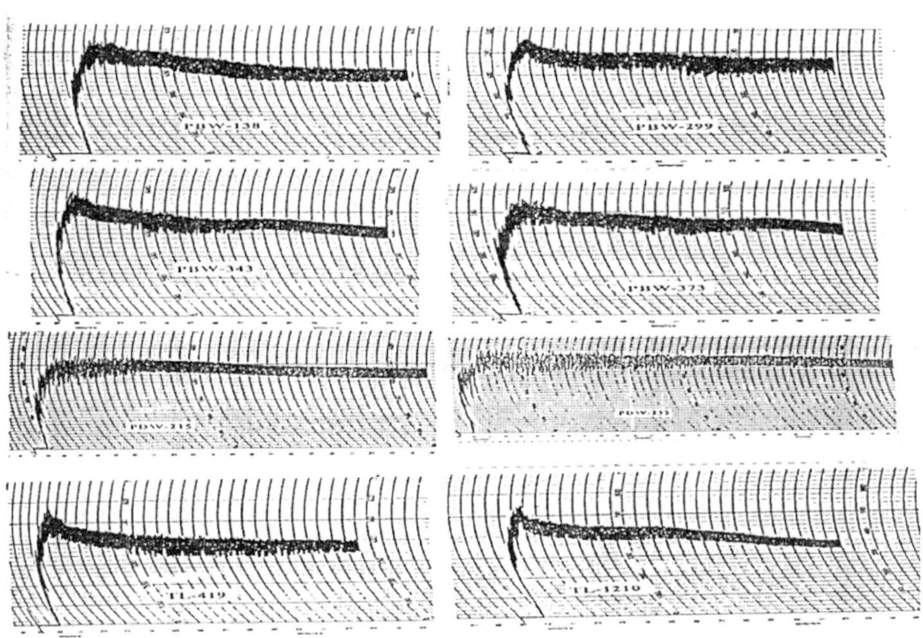

Fig. 1: Effect of variety on Farinographic properties of 70% extraction flour

Statistically bread and durum flours revealed no variation with respect to percent water absorption. Higher water absorption of bread and durum wheat varieties was the artifact of distinctly higher damaged starch content. Amongst the varieties, PBW-138 and PDW-233 recorded highest water absorption values. Like water absorption, durum wheat had highest arrival time (2.45min) dough development time (3.45min) and stability (19.65 min.) and triticale the lowest (1.1min, 1.5min and 0.95min). Bread wheat was found intermediate between the two wheat with the values (1.5min, 2.02min and 2.8 min). The mixing tolerance index (a measure of resistance of a dough to deformation) was highest in triticale flours (107.5BU) followed by bread (71.25 BU) and durum wheat (12.5BU) flours, respectively. The lowest values of mixing tolerance index in durum flour indicated its highest tolerance to over mixing. Durum wheat variety PDW-215 had highest and triticale TL- 419 the poorest resistance to over mixing. Degree of softening also revealed strong nature of durum flours (17.5BU) followed by bread (96.75 BU) and triticale (135 BU) flours, respectively.

Farinographic characteristics recorded in this study were different than those reported by Bakshi and Bains (1987), Lukow and Mcvetty (1991) and Nemeth *et al* (1994) for different wheats.

Amrylogarphic characteristics

Table 5 and Fig. 2 illustrated the pasting properties of different flours when the temperature of the system was increased gradually. Pasting characteristics of various starches in the amylograph showed large inter-varietal differences, reflecting the physico chemical differences in the makeup of the granules.

Table 5: Effect of variety of pasting properties of 70 % extraction flour

Variety	Gelatinization Temp. (°C)	Peak viscosity(BU)	Peak viscosity Temp. (°C)	Viscosity at 95°C (BU)
Bread wheat				
PBW-139	61.8	1120	95.5	920
PBW-299	62.2	771	93.7	665
PBW-343	62.8	830	92.5	645
PBW-373	62.8	925	91.3	585
MEAN	62.4	911.50	93.25	703.75
Durum wheat				
PDW-215	61.0	860	94.0	760
PDW-233	61.1	928	92.5	666
MEAN	61.05	894	93.25	7.13
Triticale				
TL-419	64.0	90	79.3	5.0
TL-1210	65.50	180	88.8	57.0
Mean	64.75	135	84.05	31.0
CD (0.05)	1.17	17.63	1.43	39.78

Data revealed statistically similar gelatinization temperatures of bread (62.45°C) and durum (61.05°C) flours compared to triticale flours (64.5°C). However significant variation was observed with respect to peak viscosity and viscosity at 95°C in the flour of different wheat. Bread wheat flour has highest peak viscosity (911.5 BU) followed by durum (894 BU) and triticale (135 BU) flours, respectively. Highest viscosity was shown by the flours of PBW-138 (1120 BU) and the lowest in the triticale flours was due to lowest falling number values reflecting the presence of excessive alpha amylase.

Varietal variation in the peak viscosity of flours indicated that the variety had great effect on the viscosity of starch. Like gelatinization temperature, bread and durum flours observed similar peak viscosity temperature of 93.25°C and triticale 84.05°C. PBW–138 revealed the highest (95.5°C) and TL- 419 the lowest (79.3°C) peak temperature at maximum point of gelatinization. Peak viscosity at 95°C recorded that triticale flour had lowest value of 31 BU compared to 703.75 BU and 713 BU of bread and durum flours. However, the values recorded in the investigation for bread and durum flour were more than those reported by Nemeth *et al.* (1994), Bakhshi and Bains (1987) but lower than those reported by Lorenz and Welsh (1977).

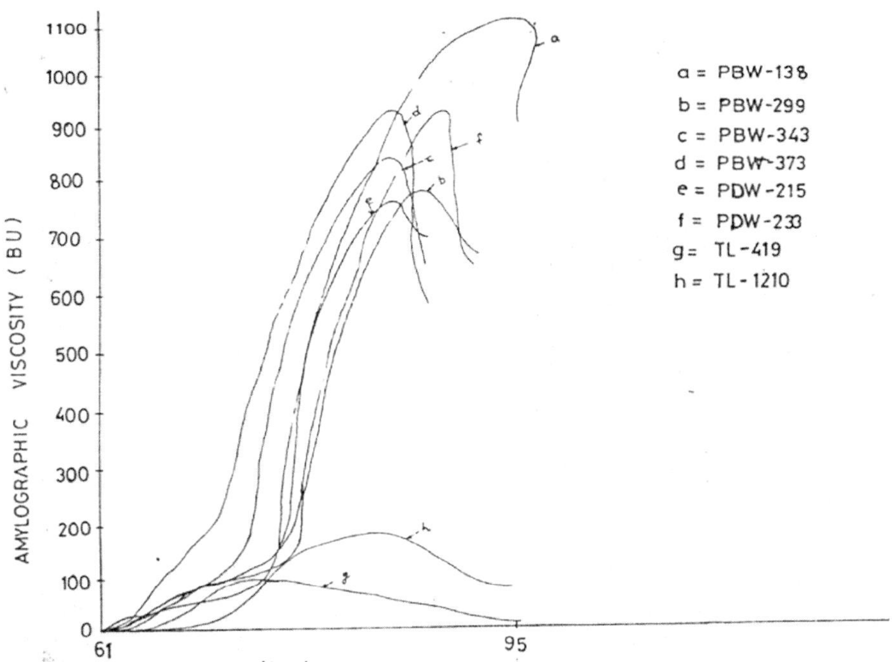

Fig. 2: Effect of variety on the pasting properties of wheat flour

Rheological studies clearly indicated that triticale flour was weakest than bread and durum flours. Triticale flour resulted in poor mixing properties, was found sticky and needed less time for mixing than standard for making pretzels.

Pretzel Production

During preparation of pretzels, slight modification of the recipe was done with respect to mixing time and water addition. Bread wheat varieties PBW–299, PBW-343 and PBW-373 were found sticky at 50% level of water addition and their threads stuck to each other thus, making separation and shape formation difficult. So water addition was reduced to 43% in PBW-299 and 45% in PBW-343 and PBW-373 flours. PBW-138 and durum varieties behaved normal as per standard recipe and method. Both the triticale varieties were found problematic with respect to pretzel production procedures. Their dough was highly sticky and breakable due to gluten dilution. So, the water level and mixing time were reduced to 43 ml, 5min and 40 ml, 4min in the flours of TL-419 and TL-1210, respectively. All other parameters for pretzel making were found satisfactory except in TL-1210 and PDW-233, which took 45min instead of 25min for drying.

Organoleptic evaluation scores indicated that pretzels from PBW-343 and PBW-373 had highest overall acceptability (Table 6). However pretzels from PBW-343 exhibited best stability during storage.

Table 6: Overall acceptability scores of Pretzels prepared from different wheat and triticale

Varieties	Overall Acceptability Scores
PBW -138	7.0
PBW-299	7.0
PBW-343	9.0
PBW-373	9.0
Mean	8.0
PDW-215	7.0
PDW-233	7.0
Mean	7.0
TL-419	7.0
TL-1210	7.0
Mean	7.0

Mean of 15 judges: 9 point hedonic scale

CONCLUSION

It was concluded that bread wheat having quality characteristics intermediate between durum wheat and triticale is best for pretzel making and varieties PBW-343 and PBW- 373 were suitable for their production at normal commercial extraction rate of flour.

REFERENCES

Anonymous (1993). Approved Laboratory Methods. American Association of Cereal Chemists, Paul, Minnesota, USA.

Bakhshi, A K and Bains G S (1987). Study of the Physico– chemical, rheological, baking and noodle quality of improved durum and bread wheat cvs. J Fd Sc. and Tech. 24 : 217-221

Bakhshi AK and Dingra D P (1991). Scope of cereal based food processing industry in Punjab. Indian Miller 21:15-19

Bhattacharya K R, Sowbhagya C M and Swamy Y M I (1972). Some physical properties of paddy and rice and their inter- relations. J Sci. Fd and Agri 23:171

Barlow K K, Butrose M S, Simmonds B H and Veble M (1973). The nature of starch protein inheritance in wheat germplasm. Cereal Chem. 50:443 .

Boyacioglu M H, Youngo V L, Khan K and Appolonia B L (1991). A comparison of durum wheat grown in Turkey and the United States. Pasta Journal 10: 24-26 .

Dexter J E, Tkachuk R and Jipples K H (1991). Physical Properties and processing quality of durum wheat fractions recovered from a specific gravity table. Cereal Chem., 68:401- 405.

Directorate of Economics and Statistics (2002). Agricultural Statistics at a glance. Ministry of Agriculture, Govt. of India, New Delhi.

Gill K S, Sehhon K S and Nanda G S (1980). Development of high yielding wheat varieties superior in bread, biscuit and chapati making. Proc. Seminar " Wheat your Needs." New Delhi pp – 189.

Henry Simon (1985). Cereal Laboratory Methods. Henry Simon Ltd . Stockport S K 30 RT , England No. 1002.

Kumar G V, Ranga Rao G C P, Venkateswara Rao G and Shurpalekar S R (1979). Variability in physico – chemical and milling characteristics of Indian Triticales. J Fd Sci Tech., 6 :184-189.

Lane, JH and Eynon, L (1923). J Sci Chem. Ind. 42:327.

Leon, A.E., Rubiolo, A and Anon, M.C. (1996). Use of triticale flour in cookies. Quality factors. Cereal Chem. 73 : 779-784.

Lorenz, K. and Welsh, J. (1977). Agronomic and baking performance of semi – dwarf triticales. Cereal Chem. 54:1049-56.

Lukow, O.M. and Mcvetty, P.B.E. (1991). Effect of cultivars and environment on quality characteristics of spring wheat. Cereal Chem., 68:597-601.

Nemeth L J, Williams P C and Bushuk W (1994). A comparative study of the quality of soft wheats from Canada, Australia and US. Cereal Fds World, 39:691-94.

Patnaik G (1992). Export potential for wheat and wheat products. *Indian Miller,* 22:31-34.

Ramandeep K (1998). Development of pretzel type products from roller milled wheat flour. M.Sc. Thesis Punjab Agricultural University, Ludhiana, India.

Sekhon K S, Saxena A K, Randhawa S K and Gill K S (1980) Use of triticale for bread, cookie and chapatti making. J Fd Sci Tech., 17:233-35.

Singh RP, Bakshi AK and Sehgal K.L. (1988) Comparative studies on baking quality of triticales, bread and durum wheat cultivars of Punjab. Indian Baker, 18:17-21.

Singh RP, Bakshi AK and Sehgal K.L and Bakshi A.K (1992). Baking quality variability in Punjab wheat. Indian Baker. 23:13-18

Singh RP, Sehgal KL, Saxena AK, Sekhon KS and Gill KS (1983). Physico-chemical, rheological and baking of characteristics of some promising strains of wheat. J. Res. Punjab Agricultural University, Ludhiana, 20:403-10.

Varughese G, Pfeiffer WH and Pena RJ (1996). Triticale: A successful alternative crop. Cereal Fds World, 41:474-482.

Venkateshawara Rao G, Ranga Rao GCP, Vatsala CN, Kumar GV and Shurpalekar SR (1977). Bread, biscuits and chapati making quality of Indian triticales. J Fd. Sci Tech. 15: 11-15.

22

Newer Products: Potential Utilization of Cereals

Savita Sharma, H.P.S. Nagi, H. Kaur and R. Kaur

Abstract

Technological advancement, increased urbanization, awareness and interest of consumer for variety products have given a great challenge to the food sector to develop inexpensive foods that are nutritionally superior and acceptable to the intended consumers. Keeping this objective in view, we have made an attempt to develop technology for variety products from wheat and corn and suitability of available varieties judged. With the emerging trends of food market towards fast foods, there are vast prospects for efficient utilization of cereals by exploring varieties products. A steamed bread (a Chinese traditional product) and Tortilla and corn chips (Mexican snacks) were developed. Steamed bread is defined as the product resulting from cooking or steaming fermented dough. Recipe and method for its preparation was standardized. Nutritionally rich steamed bread can be prepared by supplementing 10% level of chickpea and defatted soybean flour. Corn snacks have become a prime target of some consumer advocates and nutritionists who have referred them as a source of 'empty calories'. But reality is that these snacks provide instant energy, as they are rich storehouse of nutrients. Process and recipe for the preparation of Tortilla and corn chips from Indian maize were standardized. The quality and acceptability of these snacks were worked out and found that these products were highly acceptable. Indian consumers have versatile taste, so these products will be readily accepted in Indian market.

INTRODUCTION

With changing times, food habits continuously change. With economic pressure and globalization food is moving out of the kitchen. It calls for mechanization and convenience without sacrificing affordability and quality. Cities are growing at 7-12% per year. With both the parents working, ready to eat foods, and variety snacks are needed to save on time, fuel end skill. With emerging trends of food market towards fast foods there are vast prospects for efficient utilization of cereals by exploring variety foods. Cereals constitute a good proportion of the product, which have gone global. Keeping these objectives in view, we have made an attempt to develop technology for variety products from wheat and corn, which are summarized in following heads.

1. Steamed Bread

Wheat is the staple food for major segment of world population. Punjab, a major producer and contributor of wheat to the central pool is finding alternative markets in view of open trade and stiff competition from other states and countries. Although wheat production has increased tremendously over the last three decades but there has been very slow increase in the production of "value added" wheat products which is the need of the hour in order to boost profitable utilization of the surplus wheat available. Steamed bread is one such product, which can be placed on the bakery counters.

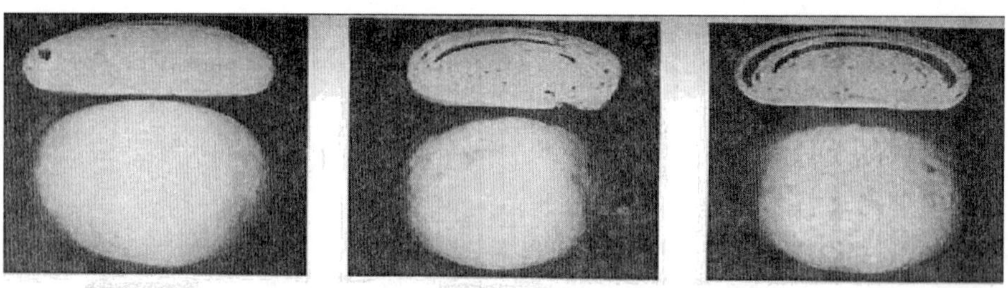

PBW 343 PDW 233 TL1210

Fig. 1: Effect of wheat varieties

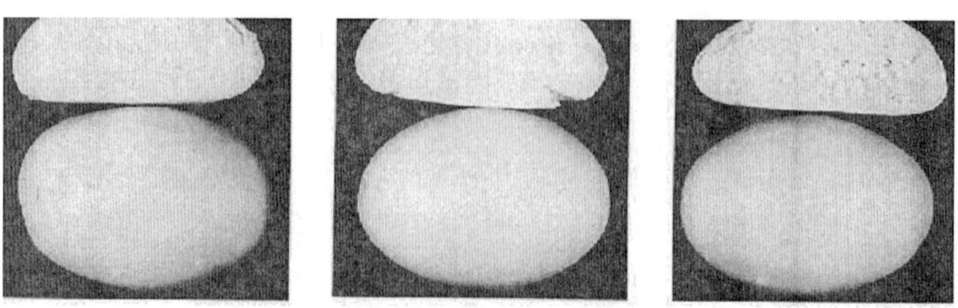

70 % 75 % 80 %

Fig. 2: Effect of extraction rates

Steamed bread is staple diet of large segment of world population and is served hot with the meals. In Northern China, nearly 70 percent of the wheat is used to prepare steamed bread and is becoming popular all over Asia. Steamed bread is defined as the product resulting from cooking or steaming fermented dough. Steamed bread does not have crust as the Western style bread. Common to all forms, characteristics sought in steamed bread are a smooth semi glossy finish, white colour inside and out, a soft moist and uniform crumb texture and a large specific volume.

Recipe and process standardized, and suitability of wheat varieties and types of flour were worked out. Among types of wheat, bread wheat (PBW-343) of 70% extraction rate flour was found to be best suited for preparation of steamed bread on the basis of volume and organoleptic evaluation. Addition of sodium stearoyl lactylate (SSL) improved the quality of steamed bread. Nutritionally rich steamed bread can be prepared by supplementing 10% level of chickpea or defatted soybean flour. Variety steamed bread had higher acceptability by the addition of spices and chopped onions.

Recommended process for preparation of steamed bread

Flour	:	250g
Water	:	Optimum
Yeast	:	5g
Sugar	:	12.5g
Salt	:	2.5g
Mixing	:	3 min
Dividing	:	100g pieces
Moulding	:	Round shape
Fermentation	:	2.5 hr
Sheeting	:	15 times
Proofing	:	30 min
Steaming	:	25 minn

2. Tortilla and Corn Chips

Snacking is being described as the problem free consumption of easy to handle, miniature portioned, hot or cold products in solid or liquid form which need little or no preparation and are intended to satisfy the occasional pangs of hunger. Snack food consumption and demand is rapidly rising because of changing food habbits, and advent of new convenient foods, in the market.

In Indian Food industry, corn has a very wide scope of its utility to 100%, out of which nearly 90% is used directly a food. A variety of snacks can be prepared from corn. These snack foods have become a prime target of some consumer advocates and nutritionists who have referred them as sources of 'empty calories'. But reality is that these snacks provide instant energy, as they are rich storehouse of nutrients. Corn snacks are rarely available in India. Due to urbanization, the life style of Indian people is also changing. The consumers are looking for new recipes and products on grocery shelves. Keeping in view, the composition of Indian maize varieties, tortilla and corn chips can be prepared from them.

Tortilla and corn chips are Mexican corn products, which involve the ancient Aztec process of nixtamalization (cooking and soaking maize in water containing lime, which softens the pericarp, endosperm and gelatinizes the starch. These chips are rich in b-vitamins and calcium content. These differ in process of production. Corn is cooked in alkaline water. The cooked, steeped corn called nixtamal is washed to remove loose pericarp and residual live. The nixtamal is ground to masa, which is kneaded: for corn chips, after extrusion, or sheeting and cutting of masa, immediate frying is done. For tortilla chips masa is sheeted and cut then baked or dried conditioned for some time and then fried.

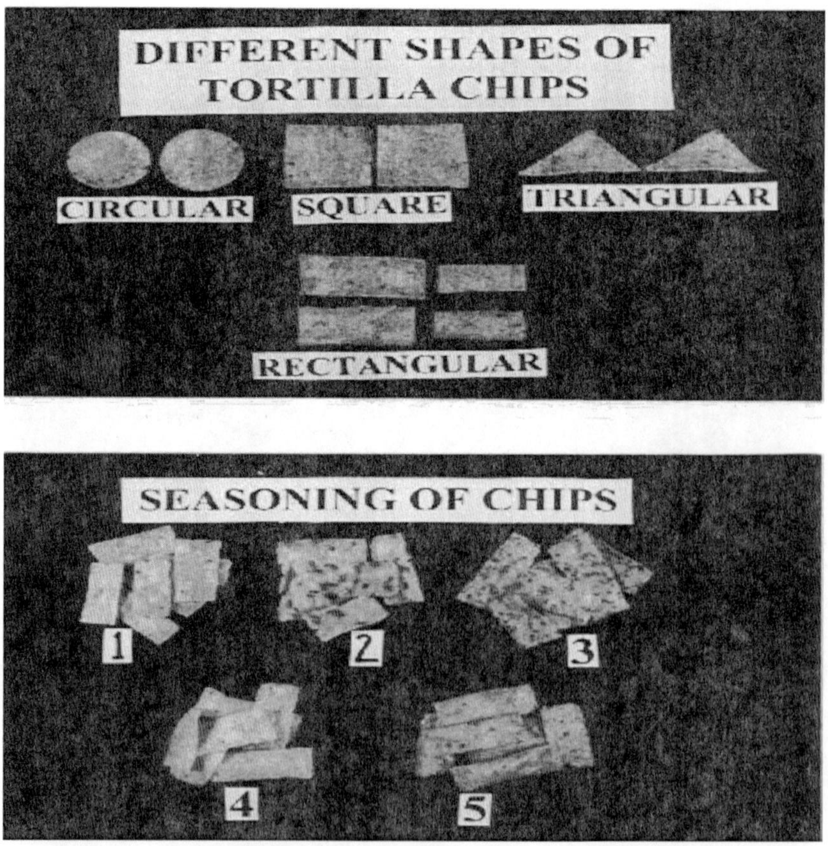

Fig. 3: Variety of Tortilla and corn chips

Preparation of tortilla chips

Corn grains + 1.5% $Ca(OH)_2$ (grain basis)
↓
300% water (grain basis)
↓
Steam cooking for 20 min
↓
Steeping/soaking for 20 hr at 30-40°C
↓
Washing with water
↓
Coarse grinding
↓
Fine grinding
↓
Masa
↓
Sheeting (0.05 cm thick)
↓
Cutting into shapes (small rectangular)
↓
Baking at 220°C for 2 min
↓
Conditioning for 10-15 min at room temperature
↓
Frying in oil at 190°C for 1.5 min
↓
Tortilla chips

An attempt was made to standardize the recipe and process for the production of Tortilla and corn chips as shown in (Fig. 3.) from Indian maize and their quality in relation to nutrition and storage studied.

Therefore these products have very bright prospects in near future in India as food markets expand new products concepts and also Indian consumers have versatile taste.

23

Reproducible Textural Characterisation of Commercial Biscuit

Bhaskar Janve, Harsimran Singh, Manoj Pandey and K. Prasad

Abstract

Three-point bend test (three point bend ring-HDP/3PB) for five types (Good day, Butter bite, Marie, Krack Jack and Salted) of commercial biscuit brands (Britannia, Supreme, Parle, Lancer, Priya Gold and Dalmia) was developed using TA-XT2, texture analyzer and probable factors affecting the results were assessed. Maximum force to break the biscuits, number of fracture before break, thickness of biscuits and percent break point penetration were found in the range of 1006.5 to 3456.5g, up to 3, 4.2 to 6.3mm and 5.54 to 22.62%. The rupture of biscuits were observed on the basis of probe penetration into the biscuits and found that maximum resistance for penetration offered by Marie whereas minimum by salted biscuits, which was respectively the hard dough and fermented dough biscuit varieties. The analysis of variance for the results revealed that there is significant difference (P£0.01) in the maximum force of break for the biscuit types and brands individually or in combination. Difference in the thickness of biscuits was also found significant among types and commercial brands. The number of fractures was not consistent among the types as well as biscuit brands before reaching the maximum force of breaking the biscuit.

INTRODUCTION

Texture is considered as a human experience developed between food structure and its response or behaviour when handled. Instrumental measure may only be utilized to quantify the physical aspects of food texture and make no allowance for the influence of factors such as physiology or psychology of perception. However, when comparisons are made like for like the data formed is invaluable as within standard quality assurance practices or in the development of new products where texture analysis becomes an integral factor in the prediction and determination of rheological characteristics. Imitative instrumental measures of mechanical texture is an attempt to simulate real-life imposed conditions in a range of applications as diverse as assessing peak force to break the biscuits. Biscuit is baked wheat product, which has found acceptability even in the rural parts of the country. The texture of biscuits is not only affected by used raw materials and manufacturing processes but also affected by the involved machinery in the line. Therefore, our objective of this paper was to investigate texture of commercially available biscuits brands, so as to correlate the manufacturing process and types with texture.

MATERIALS AND METHODS

Popular identified biscuits Good day, Butter bite, Marie, Krack Jack and Salted types of commercially biscuit brands Britannia, Supreme, Parle, Lancer, Priya Gold and Dalmia were procured from local market.

Texture measurement

Three-point bend test or snap test of biscuits was performed by using Texture Analyzer (TA-XT2), Stable Microsystems, Surrey, U.K. using three point bend ring-HDP/3PB support setup (Fig. 1).

The biscuits were placed centrally on 2-side-blade-support placed 30mm apart (Fig. 2). The upper blade test speed was 0.5mm/s. Maximum force for break the biscuit and number of fracture before final break was measured in a test run using the data acquisition (400 pps) system and analysed using Texture expert v1.22 software of Stable Microsystems, Surrey, U.K.

Thickness and Percent Break Point Penetration

The thickness was measured using Vernier calipers and the break point penetration form the force deformation graphs individually obtained from the test run. The percent break point penetration was calculated using equation no. 1.

$$Percent\ break\ point\ penetration = \frac{Break\ point\ penetration \times 100}{Thickness\ of\ biscuit} \quad (1)$$

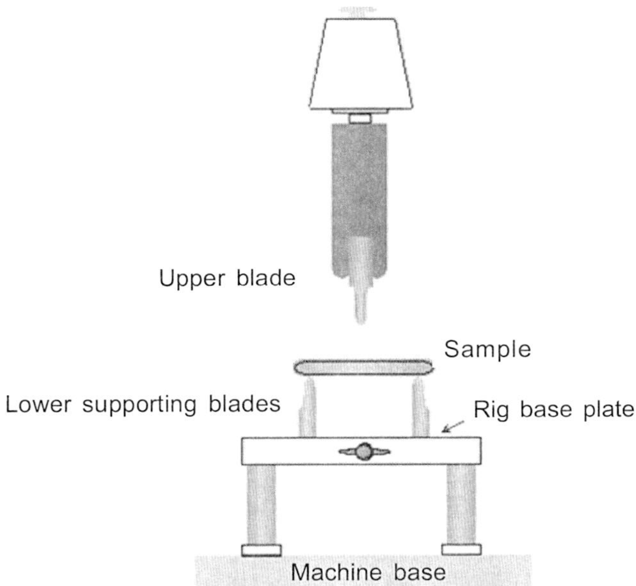

Fig.1: Support setup, 3-Point Bending Rig HDP/3P for measuring resistance of biscuit to Bend

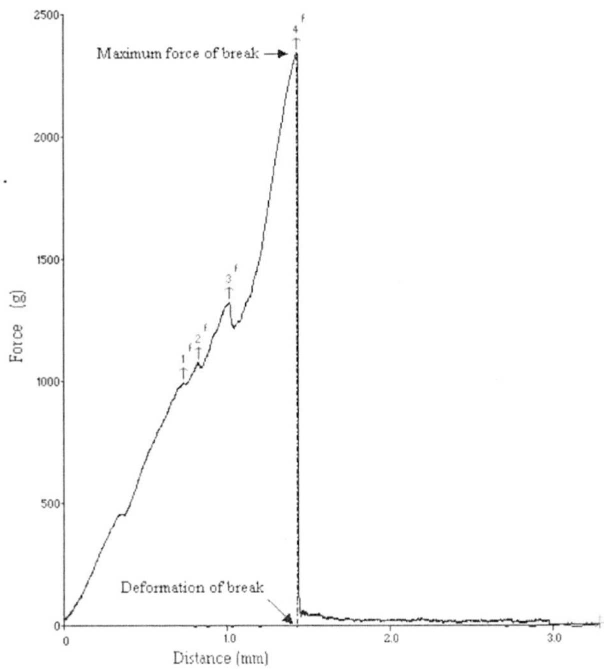

Fig. 2: Typical force-deformation cruve for a typical biscuit

Statistical analysis

The observed results were subjected to analysis of variance (ANOVA), correlation, and critical difference, at 5% level of significance ($CD_{5\%}$).

RESULTS AND DISCUSSIONS

Maximum force of break the biscuits

The maximum force required for breaking the biscuit ranged from 1006.5 to 3456.5g with an average of 2196.1g force. The analysis of variance (ANOVA) for the maximum force required for breaking the biscuit showed that there was significant difference ($P \leq 0.01$) between the biscuits types (F value 9.80), brands (F value 19.59) and as well as for the both (F value 14.64) with a critical difference of 271.70g at $P \leq 0.05$ level of significance. The maximum force required for breaking the biscuits is presented in Fig. 3, which highlights the facts.

The maximum force was needed to break the biscuits of Marie type whereas minimum force was required to break salted type. This may be due to the facts that Marie was of Semi Sweet type hard dough variety, whereas salted biscuits are cracker type fermented dough variety in which desired fermentation time is given for leavening and for characteristic crumby texture. Goodday is of Short

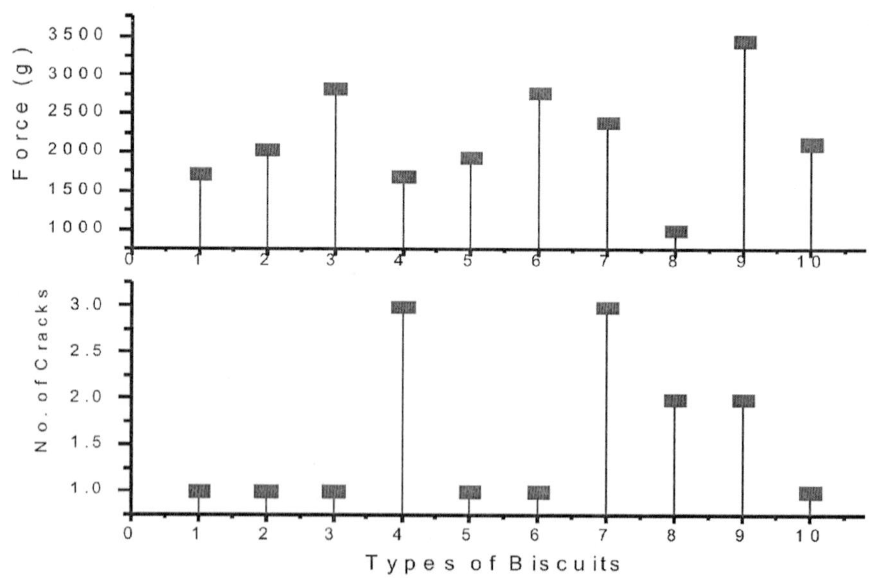

Fig. 3: Types of biscuit with (A) Maximum force to break the biscuit (B) Number of fracture before break

dough variety in which considerably high quantity of fat is incorporated in the manufacturing process where the intermediate force was found for breaking the other types of biscuit. It was may be due to the combined effect of thickness as well as the dough characteristics.

Number of fracture before break

The inconsistent number of cracks was found mainly in fermented and hard dough types of biscuits. The fractures before final breaking of biscuits were found up to 4 in number (Fig. 3). The analysis of variance (ANOVA) for the fracturability was insignificant even up to $P \leq 0.10$ among the types, brand, and both.

Thickness of biscuits

Biscuit thickness varied from 4.2 to 6.3mm, which was found that the biscuits are of significant ($P \leq 0.01$) difference in their types and brand wise but found insignificant when compared in combination with brand and types. The critical difference found at 5% level of significance was 0.65mm.

Percent break point penetration

The penetration depth before break of biscuit varied from 5.54 to 22.62% depending on the nature of biscuits manufacturing (Fig. 4). The variation in the percent break point penetration was found significant among types but not among

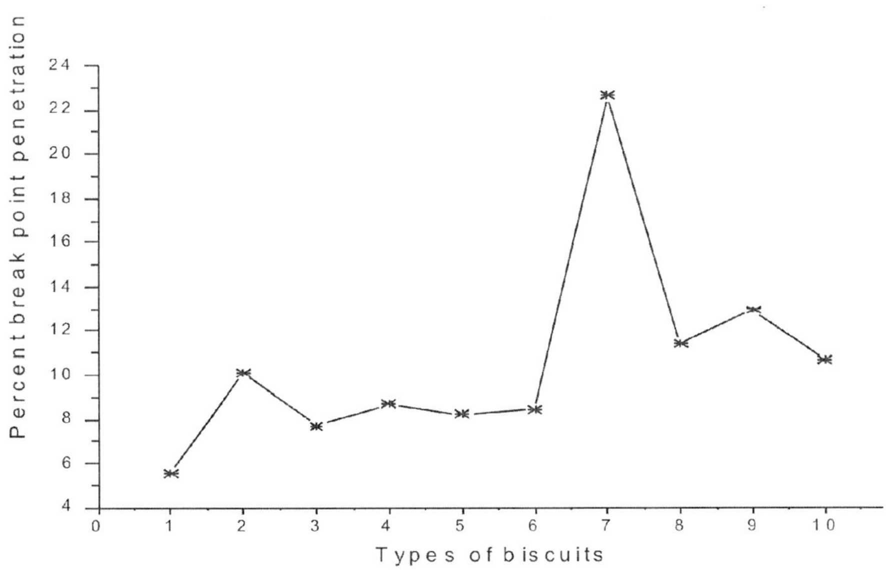

Fig. 4: Types of biscuits with peroent break point penetration

commercial brands. Although the combined effects of two parameters were found significant, this may be due to the variable total thickness of the biscuits. The maximum percent break point penetration was found for salted biscuits, which has crumby texture and fall in cracker type fermented dough variety.

CONCLUSIONS

Biscuits on the basis of thickness differ significantly among types and commercial brands. The number of fractures was not consistent among the types as well as biscuit brands before reaching the maximum force for breaking the biscuits and found that maximum resistance for penetration offered by Marie whereas minimum by salted biscuits and other biscuits are in intermediate range. Due to crumby texture the maximum percent break point penetration was found for salted biscuits.

REFERENCES

Bernussi, A.L.M., Chang, Y.K. and Bustos, F.M. 1998. Effects of production by microwave heating after conventional baking on moisture gradient and product quality of biscuits (Cookies), 75(5): 606-611.

Hoseney, R.C. 1994. Principles of Cereal Science and Technology. American Association Cereal Chemists Inc., St. Paul, Minnesota.

Kent. N.L. and A.D. Evers. 1994. Technology of Cereals. Pergamon Press, London.

Manley, Duncan, 2002. Biscuit, Cracker, and Cookie Recipes for the Food Industry, Woodhead Pub. Ltd., Cambridge.

Matz, S.A., 1969. Cereal Sciences, AVI Pub. Connecticut.

SECTION-VI
STORAGE AND HANDLING

24

An Overview on Grain Storage Management Systems

Sanjay Kumar Jain, N. Jindal and R.C. Verma

Abstract

The demand for the food for ever increasing population is always continuous, but the production is limited to specific seasons. Therefore the food grains have to be stored for sometimes. Grain is a living biological material and therefore storage structures should be suitably designed to protect them from the ravages of pests, rodents and micro-organisms. However, the storage losses account for a major portion which can be reduced to considerable extent if proper care is done. The faulty design, poor maintenance and limited knowledge of the farmers concerning storage technology have greatly added to the spoilage of the food grains during storage. Sometimes the foods if not properly stored may be edible but it may lose the essential vitamins and minerals, thus their nutritive value has been lost. There is a major requirement of a good storage structure. Various food grain storage structures (small and large scale) are described.

INTRODUCTION

The technological advancement in agriculture and its allied fields are greatly responsible for higher food production in our country. A record high figure of 212 million tonnes of food grains were produced in 2002-2003 as compared to 55 million tonnes in 1950-51. The demand for the food for ever increasing population is always continuous, but the production is limited to specific seasons. Food Corporation of India and other government agencies procure and handle a very small percentage of this huge production and most of it remains with farmers and traders. According to one estimate about 70% of the total food production is retained by the rural farmers to meet their food and seed requirements and rest enters the urban market channel and public distribution system. Therefore the food grains have to be stored for sometime. The period may vary from few weeks or few months to few years, depending upon the requirements. A sizable quantity of about 10-20% of the retained food grain is lost due to improper handling and inadequate storage conditions in Indian conditions.

Grain is a living biological material and therefore storage structures should be suitably designed to protect them from the ravages of pests, rodents and micro-organisms. However, Food and Agricultural Organization have estimated such losses may be as high as 20 to 30% in developing countries under tropical climates. Although the grain losses occur at different stages beginning from farmers' field till it reaches to the final consumer's dining table. However, the storage losses account for a major portion of such losses. These losses can be reduced to considerable extent if proper care is to be done. However farmers are not adequately equipped and trained to store grains safely for more than few weeks. Farmers prefer to sell their produce immediately after threshing the crops due to unfavorable circumstances. Even if they store the grains, they have a variety of problems in their storage system and huge quantity of food grains is lost. There are many factors responsible for grain spoilage; pests and rodents are the primary cause especially in the region of high humidity and high temperature. The faulty design, poor maintenance and limited knowledge of the farmers concerning storage technology have greatly added to the spoilage of the food grain during storage. Sometimes the foods if not properly stored may be edible but it may loose the essential vitamins and minerals, thus their nutritive value has been lost. They may become toxic even.

The storage structures used in different parts of the country are of numerous types and there are lots of variation in their materials, constructional methods and capacities. These storage structures are not technically sound and modifications are highly desirable. The time has come to keep down the food grain losses and wastage at minimum to feed the ever increasing population and it is extremely important in Indian conditions where cultivable land is decreasing at very fast pace.

The major requirement of a good storage structure can be enumerated as follows:

1. The storage structure should be moisture proof.
2. The storage structure should be rodent and termite proof.
3. It should have low thermal conductivity.

4. It should provide air tightness.
5. It should be equipped with easy loading and emptying mechanism.
6. It should be cheap and should involve the local material of construction.
7. It should not cause the loss of structure of grain by dehydration (wilting, shriveling).

However, all of the above parameters are not fulfilled in the rural farm storage.

Losses of Food Grain

The qualitative and quantitative losses in food grains may occur separately or together. The chemical changes occurring in the protein and the carbohydrates, the contamination taking place due to chemical toxins, insect's fragments and rodents' urine and excreta are mainly responsible for the loss in quality of the food grain in the storage. In storage, both the losses occur simultaneously. The seed loss is another major loss and there are many factors responsible for this. The viability of seed may reduce if seed coat is damaged during filling and emptying the storage structures. The chemicals used to control the insect infestation also affect the seed germination. The viability of seed is also affected by the development of microorganisms. The chemicals cause abnormal growth of root and shoot, reduction in vigour and other undesirable physiological changes.

The Table 1 shows the factors influencing the storage losses. However, moisture, temperature, insect and rodents and conditions of the storage are the predominating factors.

Table 1: Factors affecting losses in grain storage

S. No.	Factors
1.	Moisture
2.	Temperature
3.	Insects and rodents
4.	Quality before storing the grain
5.	Type of storage bin
6.	Sanitation
7.	Use of pesticides and fumigants
8.	Mechanical loss factors
9.	General condition of location of storage.

Grain Moisture

The moisture content is the single most important factor responsible for the deterioration in the quality of the food grain in the storage. It is often said that reduction in 1% of moisture content will double the shelf life. The amount of moisture in grain determines whether mould damage will occur, even at freezing temperatures. If there is sufficient moisture for mold growth, the temperature determines which mold growth will grow and how fast they will develop. In any case higher the moisture and the temperature, the faster the growth and more rapid the spoilage of grain. However, when the moisture content is low enough, all mold growth stops even at favourable temperatures. There is a certain limit of grain moisture of different commodities in relation to prevailing conditions of environment that should be kept in mind before storing the grains. The moisture to which grain will dry is determined by the relative humidity of the surrounding atmosphere. This property of grains is referred as equilibrium moisture content. The maximum level of moisture content for safe storage is given in Table 2.

Table 2: EMC of various grains at 70% RH

Commodity	Equilibrium Moisture Content at 70% RH
Paddy	15.0
Rice	13.0
Maize	13.5
Wheat	13.5
Sorghum	13.5
Cotton seed	10.0
Ground nut (shelled)	7.0
Copra	7.0

The Table 3 shows the safe storage moisture according to storage period for cereals. These limits of moisture have been fairly well established through research and experience.

Table 3: Safe storage moisture for various food grains for 1 and 5 years

Commodity	Moisture required for safe storage% (w.b.)	
	for 1 year	for 5 year
Barley	13.0	11.0
Corn(Maize)	13.0	10-11.0
Oats	14.0	11.0
Rice	12-14.0	10-12.0
Rye	13.0	11.0
Sorghum	12-13.0	10-11.0
Wheat	13-14.0	11-12.0

The moisture is also an important factor in the activity of insects. It has been found that when the moisture content is as low as 9% in shelled corn and wheat, most insects become inactive. While such low moisture content is not readily obtainable in current practices nor desirable because of shrinkage; experience shows that the drier the grain, the less the insect hazard. Insect damage in stored rice below 12.5% moisture content is much less than with moisture above this level.

Temperature

The temperature controls the activities of the microorganisms and its effect is also correlated to the moisture present in the grain. When grain in storage is cool, there is little likelihood of spoilage. Low temperature offsets the effects of high moisture, particularly as it affects the development of molds and insects. One of the greatest benefits of moving and turning grain, so widely used in elevators, is cooling. This, however, can be accomplished more effectively by aeration which is used to cool and hold damp grain in storage for weeks or even months. The respiration rate of grains is low at low temperature. Under favourable conditions, sometimes the respiration rate reaches so high that the germination occurs. These chances are even high in the freshly harvested grains with high moisture if stored without aeration. The grains contain initial internal infestation as well as field and storage fungi. Since the infestation and infection by storage pests and organisms are inherent, the atmosphere inside the storage should be capable of depletion of oxygen and moisture migration. The temperature gradient and moisture migration are generally present in the metallic structure with high thermal conductivity. This may induce the initiation of dew point and condensation of moisture within the bulk. This condensed moisture increases the higher moisture pockets. It may be responsible for high rate of respiration, mould growth and threatened germination. If a light roof coloured with white paint over the top (which reflects the solar heat) is provided then the heat transfer can be minimized.

The insects accelerate their growth up to 42°C but fungi have tolerance limit from 20°C to a maximum 63°C. Therefore these environments have to be avoided for safe storage requirements.

Fungi, Mites, Insects and Rodents

The micro- organisms are omnipresent. They may come from soil, from growing plants, on the surface of container and in the air. There are two types of micro-organisms: the budding and fission fungi. Fungi are the most dangerous and extremely harmful enemy for the grain as they cannot be seen with naked eyes and their harmful influence spreads very quickly. They consume the soluble constituents of the food tissue and may generate organic compounds that make food unpalatable. On the other hand the presence of rodents is conspicuous by the holes. Most storage fungi grow to a maximum in the temperature range of 20°C to 40°C under high humidity conditions. However, growth of some of the species is restricted under these conditions. Some fungi produce chemicals which are toxic to human and domestic animals. The growth of these fungi is more in the coastal areas of the country where high humidity and favourable temperature exist.

Mites are extremely small having a shining translucent body and legs. They vary in colour from pale straw to dark reddish brown. Mites are usually seen if they are large in numbers and are visible as dust on the surface of the bags. The damage from the insects is the principal loss in stored grains in the coastal areas of the country where the climate is warm and humid. The insects may also cause the problem in low temperature areas if the moisture content is high and these are sources of infestation. The undesirable insects/pests causing deterioration and loss to stored grains are mainly beetles and moths. They do not breed successfully if the relative humidity of the environment is less than 40%. Cereals are vulnerable to rodent attacks and damage of this effect is maximum at the village level farm storage. Rodents carry diseases transmissible to human and are also a source of contamination within the stored grains.

In any case the quality of the grains before storing in the bin plays an important role. If infested grain is stored in the bin, there is high probability that the infestation will grow. At the same time infested bins also accelerate the contamination of the grain during storage. It is therefore, desirable to store uncontaminated clean grain in insect free and sound storage structures of proven type.

Grain Storage Structures

The preliminary storage of food grains begins after the crop is harvested. In India the freshly harvested crops is heaped on the threshing floor and remain there for quite some time before threshing. However, it should not be kept for more than a day or two without drying. However, these heaps can be stored for longer period if harvesting conditions coincides with the dry weather which to some extent helps in the drying. Hereafter the threshed grain is stored either in indoor or outdoor structures. The indoor storage structures are employed by the small and marginal farmers with the main objective of storing the grain both for food and seed. Outdoor structures are seldom used by the small farmers. Storage can also be distinguished as (1) underground and (2) above ground.

Under Ground Storage Structures

This type of storage structure found their place in low water table tracts especially of the dry parts of Karnataka, Uttar Pradesh, Bihar, Madhya Pradesh, Maharashtra, and Andhra Pradesh. The safety and ease in maintenance are the prime factors responsible for their popularity. They are rectangular or conical in shape and have a narrow mouth to load the grains. These types of structures are easy to construct as only pits are to be dug up and lined with straw to absorb the moisture. After filling the food grains, the inlet is covered by the stone slab. Infestation is generally low due to depletion of the oxygen but seepage of moisture may result in the mould growth. The rodents are also a problem in these types of structures. The loading of the structure is a problem in these types of structures. The fumigation is also a problem. The peripheral layer of grains is spoiled generally due to seepage from the walls or bottom of underground structure. These grains become dull in colour and mushy in odour. The viability of seed is also affected. The food sometime may not be suitable for human consumption. The CFTRI has improved these type of structures by giving a polyethylene lining on the walls, periphery and at the floor with bitumen which reduces the ingress of the moisture.

Above Ground Storage Structure

Bag storage: Generally the capacity of gunny bags is 50 or 100kg. This type of structure is simple in construction and low in cost. The traders and processors store grains for trading purposes and they seldom worry about the quality of the food. The problem of deterioration exists in all types of storage structures however, the degree of damage varies.

Bukhari: Bukhari type structures of varying shapes, size and material of construction are used in different parts of Assam, West Bengal, Orissa, Bihar and Madhya Pradesh. The structure is made of split bamboo-strips with a bottom of either made of bamboo or wooden planks. It is mostly cylindrical in shape. The structure is generally used for storing seed like wheat, gram, maize and jawar. The structure is also used for storing wheat straw for cattle feed. About 10-15 quintals of seed are stored in bag storage form, although the space available can accommodate a lot more than this quantity. A layer of wheat straw about 150cm thick is first spread at the bottom before placing the seed bags in stack. Finally the bags are covered with straw till those remain in the storage. The outlet in the lower portion is not necessarily for taking out grains. It is helpful in overall cleaning of structure. The general defects like inadequate height of the plate form, unsuitability of structure for fumigation, not being moisture and fire proof and vulnerable to insects and rodents have been minimized in the improved Bukhari type structures. An outlet is also provided at the bottom of structure. Some details of various storage structures are given in the Table 4.

Oil and coal tar drum storage structure: Abandoned drums of oil and coal tar are utilized by some villagers to keep sizeable quantity of their farm produce. These are readily and economically available in town and cities. The capacity of 220 litres drum is about 175 kg of grain. As improvement can be made by providing lid half open and using some metal primer after thoroughly cleaning the container. These bins are reasonably cheap, durable and suitable for grain storage. It would be desirable to empty the drum be carefully cutting the top portion of the cylinder so that drums could be later modified as storage structures.

Morai: Morai is an outdoor structure exclusively made of paddy straw ropes. Sometimes mud is also used for plastering the bottom portion of the structure. There are different local names prevalent for the same structure. In Andra Pradesh it is called Puri, Mora is the term used for the same in north Bihar and Oliya in Orissa. The holding capacity of the structure also differs from location to location and according to need of the owner. In West Bengal its capacity ranges from 2 to 5 tons. It is constructed over a hard and elevated surface utilizing 10-15cm diameter straw rope which is wound to form a concentric circular wall. Simultaneous loading and winding operations are performed. Sometimes metallic base wall is provided as apart of Morai and is very useful as it provides additional support to the structure. The brick masonry base constitutes the additional cost of structure. The structure is susceptible to fire. Neither it is air tight nor moisture proof. Insects and rodent can cause sizable damage to the stored grains.

Table 4: Details of various food grain storage structures

S. No.	Type of bin	Material of construction	Capacity (quintal)	Expected life (years)	Whether suitable for fumigation
1.	Untreated bamboo bin	Bamboo strips	2	1	no
2.	Clay and cow dung plastered bamboo bin	Bamboo clay and cow dung	2	3	no
3.	Bamboo cement bin	Bamboo, cement and sand	2	10	yes
4.	Bamboo linseed oil and ash bin	Bamboo, linseed oil, ash and cow dung	2	12	yes
5.	Bamboo reinforced concrete bin	Bamboo cement concrete	3.25	35	yes
6.	Ferro cement bin	Wire mesh, cement, sand and mortar	1.5	40	yes
7.	Steel reinforced bin	Steel reinforced cement concrete	3.25	40	yes
8.	RCC ring bin	Steel reinforced cement concrete	10	40	yes
9.	Pucca kothi	Brick, cement and mortar	10	40	yes
10.	Metal bin	Galvanized Iron sheet	10	35	yes
11.	Bomboo mud plastered polythene lined bin	Bamboo, clay, cow dung, polythene	1.6	3	yes
12.	Pusa bin	Kachha bricks. Cow dung, polythene	3.0	12	yes

Large Scale Storage Structures

The storage and handling of food grains by various agencies like CWC (Central Warehousing Corporation), SWC (State Warehousing Corporation), FCI (Food Corporation of India), Government Civil Supplies Department, have assumed greater importance nowadays. The number of godowns and warehouses has been built to store the food grains.

CAP: Nowadays, the Food Corporation of India has to handle the large quantity of the grains and it did not had facilities of their storage properly, therefore a new concept of CAP was initiated. An innovative technology Cover and Plinth (CAP) was introduced to overcome the lack of storage space temporarily. It got such popularity that it has become a permanent system of storage in many warehouses of FCI, CWC and SWC's. They are made up of heavy duty LDPE film. It is an improved structure for storing the grains in open on the plinth which are essentially a wooden dunnage having damp and rat proof arrangements. The grain bags are stacked to standard size like any commercial storage space. Previously each stack of grains are covered by the polyethylene sheet or by the tarpaulins to protect the grains, but nowadays grain stacks are covered by the 1000 gauge LDPE sheet. Wheat, maize, gram, paddy, and jowar are mostly suited to CAP storage facilities. A CAP cover weighing 50 kg can store 150 to 200 MT of food grain for 1 to 1.5 years. This way the CAP covers are the most economical of all the storage methods adopted.

Silo storage: The bulk grain storage is now gaining popularity especially where grains to be handled are in a big quantity. The metallic and cement concrete structures have limitations regarding the diurnal temperature changes. The fungal attack is a big problem in these types of structures due to development of thermal gradient, moisture migration and localization of moisture pockets. Apart from this due to 30% void spaces present in the storage structures, they cannot provide the hermitic condition to inhibit insect growth. Once the growth of insects and microbial activity starts in one pocket, it catalyzes the deteriorative activity from one focus to other. In order to overcome these difficulties, aeration arrangement or turning the grains is absolutely necessary. This requires pneumatic or other mechanical feeding and discharging along with aeration and fumigation devices. These portable equipments will be economical only if minimum size of the storage is beyond 5000 tonnes and turn over in a year should be at least 3 times its capacity.

Pusa bin: IARI, New Delhi has improved the conventional grain storage structures by inserting a LDPE film between two layers of mud walls. It is termed as Pusa Bin. The LDPE provides the moisture proof arrangement and the mud walls keep the stored grains free from effect of temperature changes. This bin has been successfully used for the various types of food grains and it has been reported that the seed stored in this type of storage structure retains their quality over along period of time. The bin is easy and cheaper to construct as compared to other scientific grain storage structure. The Pusa bin provides damp proof structure as well as it limits the permeability of gases. It has low thermal conductivity and it has resistance to rodent attack.

Chemical Control of Insects and Rodents in Stored Grain

In India insects, microorganisms (fungus), and rodents cause heavy loss to the food grain. Ravages of such pests are most noticeable in rural household, warehouses and on farmlands. In most cases, the use of pesticides is the only reliable method of destroying them and thus preventing these losses. As most of the rural based storage structures are not scientifically pest proof, the application of pesticides is highly advocated. There are number of factors, which must be borne in mind before planning the spray or fumigation of the stored grain both for large, and small-scale storage system. The factors are as follows:

1. Type of cereal being stored;
2. The type of pest present or likely to attack;
3. Type of storage structure being used;
4. Projected storage period;
5. Chemicals available and expertise to handle it;
6. Ultimate use of the grain either as seed of animal feed or for human consumption.

As far as possible chemicals should be used as fumigant for control of established infestation and contact insecticides to prevent reinfestation. The Table 5 shows the list of some of the common insects, pests, and rodents found in stored grains.

Table 5: Some common insects, pests and rodents of stored grains

Local name	Scientific name
Rice weevil	***Sitophilus oryzae***
Granary weevil	Sitophilis granarius
Lesser grain borer	Rhizopertha dominica
Khapra beetle	Trogoderma granarium
Pulse beetle	Collosobruchus spp
Red flour beetle	Tribolium cost anem
Confused flour beetle	Tribolium confusum
Saw toothed grain beetle	Oryzaephilus surinamensis
Grain moth	Sitotraga Cerealella
Rice moth	Corcyra Cerealella
Fig moth	Ephestia Cautella
House mouse	Mus Musculus
Black rat	Rattus rattus
Brown rat or Norway rat	Rattus Poryegicus
Lesser bandicoot	bandicoot bengal ensis
Larger bandicoot	bandicoot Indica

Once the factors have been carefully looked into one may conclude that following definite characteristics are required in the fumigant:

1. A fumigant should be highly toxic at all stages of insects
2. It should have high vapour pressure and have no adverse reaction with the stored grains.
3. It should be non-inflammable and devoid of fainting.
4. It should not settle down immediately after application.

5. It should not affect the viability of treated seed and be known antidote and treatment
6. It should be indigenously manufactured and be economical.

The use of chemicals would be effective, only if applied in proper dose to cover particular area or volume. The Table 6 shows the application rate of various chemicals.

Table 6: Recommended dosage of widely used chemicals for stored grain

Chemical	Dosages
Malathion	Dust1. 0.250kg per 10 m^2 of surface area2. 0.120kg per 10 m^2 of surface areaSpray1. Surface treatment: 3litre per 300 m^3 at an interval of three weeks2. Air charging: 1litre per 300 m^3 at an interval of not less than seven days (two-three weeks)3. Walls and alleyways: 3litre per 100 m^3 once a week in order to control crawling insects
Aluminum Phosphide	100 to 140 tablets per 100 m^3
Methyl Bromide	1.5 to 2.2 kg per 100 m^3
Benezene Hexochloride dust containing 0.65 % Gamma isomers	100 g per 4 m^2 of surface
Carbon disulphide	2kg per 30 m^3 space holding about 20 tonnes of stored grains
Lindane (Gamma BHC)	225 g of 0.04% lindane per 45 kg of grain
Pyrethrum	0.5 kg pyrethrum per 90 kg of maize

In addition to the chemical methods, farmers use some traditional methods to destroy insects. Conditioning in the sun at a regular intervals is one of them. Insects feel uneasy at a temperature above 40-44°C and try to run away from the grain mass. However, conditioning in sun does not kill the eggs and larvae inside in also carried out. Neem (*Azadirachta indica Juss*) leaves and ash are commonly used in stored grain in Bihar and West Bengal. Sometime onion bulbs relatively of smaller size are placed in the stored grain. These admixtures work, as repellant to insects. Wood ash and chopped wheat straw are also used especially when grain is to be used for seed purposes. By adding straw or ash, insects are deprived of oxygen and also feel uneasy in crawling. Smoking is also used as one of the measures to check insect infestation.

Code of Practices for Safe Storage of Grains at Farmers Level

The traditional methods practiced by the farmers with regard to the storage of food grain and seed have evolved mainly on the basis of their past experience passed on to them by their forefathers. Such practices are not always scientific and may not be considered safe to be followed for preventing grain losses during storage. Based on the scientific principles and successful results obtained in the past, the following code of practices may be recommended to the small and marginal farmers. There may be different opinions for sequence of practices but above all, practices followed have the same objective and that is to store grain without loss. It is also pointed out that the extension workers, the traders and other involved in the grain storage programmes may also take note of the practices in making their programme a success:

1. As far as possible, clean the grain prior to storage.
2. Dry the grain either in the sun or in the dryer till moisture is down to the safe limit.
3. If maize is to be stored, put kernels into the structure right after removing them from the cob. Maize should not be exposed to prevailing weather, once it has been dried.
4. Clean the storage structure and disinfect it before filling the grain.
5. Fill the structure up to the capacity carefully always from the top.
6. Stir grain with large stick so that grain settles.
7. Add recommended doses of insecticide in to the grain before closing the cover of the structure.
8. Close inlet and outlet covers immediately as flying insects may enter the stored grains.
9. Put mud or other binding material as the case may be on cracks and crevice. This would not allow the insects and pest to enter the grain or structure.
10. Area around the structure must be cleaned to discourage rats hiding around. Rats do not like clean premises.
11. Inspect structure regularly to detect creaks etc. Plaster the cracks after disinfecting.
12. Keep the outlet cover locked in the outdoor structure to avoid any risk of pilferage.
13. Spilled grain around structure should be removed, else insects and rats would be attracted to the structure.
14. The cover should be firmly plastered after removing the grain or seed from the structure.
15. No grain should be left in the structure while emptying, otherwise it would attract undesirable insects and pests.
16. Sweep all the debris and dust etc and disinfect the structure.

17. The structure should be thoroughly cleaned and made free from insects before filling the next lot. Sometimes a small fire sufficient to create smoke is advisable in case the structure is fire proof.
18. Once the structure is repaired, cleaned and disinfected, it is ready for the next filling.
19. It is always desirable to consult the local agencies regarding use of chemicals and improved techniques for grain storage.

REFERENCES

Harvest and Post Harvest Technology Schemes, Annual Reports
Indian Grain Management Research Institutes, Annual Reports

25

Management to Maintain Stored Grain Quality

Deepa Srivastava, Tanuja Srivastava, Sukhcharn Singh, C.S. Riar and D.C. Saxena

Abstract

Maintaining grain in good condition in storage requires careful, routine inspection and good storage practices. An integrated approach of several management tools and practice factors are necessary to manage stored grain to prevent deterioration and wastage: This includes: 1) harvesting, handling, and storage equipment; 2) initial condition of the grain; 3) grain moisture content; 4) grain temperature; 5) aeration system management ; 6) insect and mold control; and 7) monitoring grain condition. An effective management program greatly enhances the likelihood of maintaining top quality grain. A well- designed and properly operated storage system with adequate aeration capacity, storing only clean grain at the proper moisture content and temperature, sanitizing the bin before loading, checking the grain condition regularly and correcting problems before they get out of hand are critical management strategies that must be implemented to prevent grain damage and possible serious economic loss.

INTRODUCTION

In recent several years, the number of stored grain loss during the storage period in the world has been increasing due to the very poor or no scientific management of stored grain, especially in most of developing countries. Thus, it is really necessary to construct an advanced management system of stored grains with some current advanced techniques of information science for implementing their scientific management in order to increase modern level of grain production. In fact, because of the possible large size of grain shortage in the world extent in next century, the constructions of advanced stored grain facilities and related management systems will become a more and more problem in the future. However, how to design and construct the management system has been being a very complex and important problem in the researches on stored grain management engineering.

In past many years, a number of technical papers had discussed deeply the problems how to design and implement a number of stored grain management systems and provided many usual practical experiences and outcomes. But, we consider that those outcomes lack a scientific system method to describe and solve this problem, so that so far is how to implement the scientific management of stored grain still a very complex and difficult problem. In general, this stored grain management belongs to a problem of large complex system management about optimal control and decision that involves other problems about how to control and manage many complex objectives. Thus, there are a number of factors to influence the design and implementation of this system, which include the stored grain kind and quality, stored grain facilities, environmental temperature and moisture, and various possible pests, etc. Main objective of this management is to control the stored grain environment using natural and artificial aeration means to destroy any possible conditions resulting in any possible stored grain disasters in order to assure that temperature and moisture of stored grain are in the safety range. In general, an advanced stored grain management system based on a green control rule (no chemical means) must complete two basic tasks as follows.

1. Estimate or predict the temperature and moisture values of any position in a stored grain entity in any necessary time.
2. Manage or control various possible dangerous states of the temperature and moisture in the entity for storing grain because they could result in various possible stored grain disasters.

When stored grain goes out of condition, it is usually the cumulative result of several interrelated management factors that include: 1) harvesting, handling, and storage equipment; 2) initial condition of the grain; 3) grain moisture content; 4) grain temperature; 5) aeration system management; 6) insect and mold control; and 7) monitoring grain condition. An effective management program greatly enhances the likelihood of maintaining top quality grain.

Harvesting, Handling and Storage Equipment

Grain bin and equipment maintenance prior to and during harvest lays the foundation for ensuring that good quality grain goes into storage and that grain quality is preserved.

Harvesting Equipment

Remove all traces of old grain from combines, truck beds, grain carts, augers, and any other equipment used for harvesting, transporting, or handling grain. Even small amounts of moldy or insect-infested grain left in equipment can contaminate a bin of new grain.

Adjust combines according to the manufacturer's specifications to minimize grain damage and to maximize removal of fines and other foreign material. Most manufacturers have instructional materials and/or videotapes to demonstrate effective combine adjustment and performance.

Operate augers at full capacity to reduce wear and grain breakage. With variable incoming flow rates, reducing auger speed can keep the auger operating at full capacity. Another option is to add a hopper over the auger intake, keeping it full. Be sure all safety shields and intake grates are kept in place and in good working order.

Bins and Other System Components

Check the bin site, and remove any items or debris that would interfere with safe, unobstructed movement around the bin. Remove any spilled grain and mow the site to reduce the likelihood of insect or rodent infestation.

Inspect bins and foundations for structural problems. Uneven foundation settlement can cause gaps at the bottom edge of the bin. This can result in grain spills and provide entry points for water, insects, and rodents. Gaps between the foundation and bin will allow air that normally would be forced through the grain to escape from the bin. Small gaps usually can be filled with a high-quality sealing compound. Be sure all anchor bolts are tight and undamaged.

Inspect the bin roof and sides, inside and out, for leaks, loose or sheared bolts, and rust or other corrosion. Check the roof vents and access hatch, and caulk any cracks at the roof line. Be sure the access ladder is complete and securely fastened to the bin. Repair or replace any deteriorated parts.

Wiring for fans and other electrical components should be inspected for corrosion and cracked, frayed, or broken insulation. Exposed wiring should be routed through waterproof, dust-tight conduit. Avoid kinking the conduit, and make sure all connections are secure.

Check fans, heaters, transitions, and ducts for corrosion, and remove any accumulated dust and dirt that will reduce the operating efficiency. Be sure all connections are tight and shields are in place.

Thoroughly clean the bins. Use brooms and vacuum cleaners to remove all old grain. Do not put new grain on top of old grain, as this is a potential source of mold and insect infestation of the new grain. Also clean bins not being used for storage to keep insects from migrating to other bins. Although it is difficult to thoroughly clean under perforated floors, a grain vacuum can be used to remove much of the accumulated dust and fine material. Removing the fan allows access to the underside of the floor.

If long-term storage (over 10 months) is planned, consider treating the bin with an approved insecticide two to three weeks before new grain is added. Treat as many surfaces as possible, especially joints, seams, cracks, ledges, and corners. Spray surfaces to the point of runoff. Also treat outside the bin at the foundation and around doors, ducts, and fans. As with all pesticides, read and follow label directions during handling, mixing, application, and clean-up.

Initial Grain Condition

Good quality, clean, sound grain is much easier to maintain in storage than cracked and broken grain. Broken kernels will mold three to four times faster than whole kernels. Broken grain also is more susceptible to insect attack, since some insects feed only on broken or cracked kernels. It is critical that initial grain condition receives considerable attention.

In addition to careful combine adjustment and operation, moisture content is a key factor influencing grain damage during harvest. The least grain damage generally occurs at moisture contents of 18 to 20 percent. The amount of damage increases slightly for moisture contents below this range, but increases rapidly for moisture contents above this range. Also, frozen grain damages easily, and combining or handling should be avoided if possible.

Grain Cleaners

In addition to increased mold and insect activity, fines and broken or cracked kernels disrupt airflow distribution, contributing to storage problems. Typically this occurs in the center of the bin. The best solution is to use a grain cleaner to remove the fines before placing grain in storage. Typical grain cleaners are gravity screens, perforated augers, and rotary screens.

Gravity screens pass the grain over a screen during the handling process. Some are stand-alone units fed from an auger. Others are part of a permanent installation such as at the output of a bucket elevator. Perforated augers have a section of auger housing with multiple, small holes. As the grain is conveyed over this section of the auger, the fines are separated. This type of cleaner has limited effectiveness when the auger is operated at full capacity. Unfortunately, operating the auger at less than full capacity results in increased grain damage. Rotary screen cleaners use a rotating screen that tumbles and separates the fines and foreign material from the grain as shown in Fig. 1. These can be very effective if operated according to the manufacturer's recommendations. Grain cleaner capacity typically ranges from 500 to 3,500 bushels per hour. These capacities are approximately equivalent to 4 and 10-inch augers, respectively.

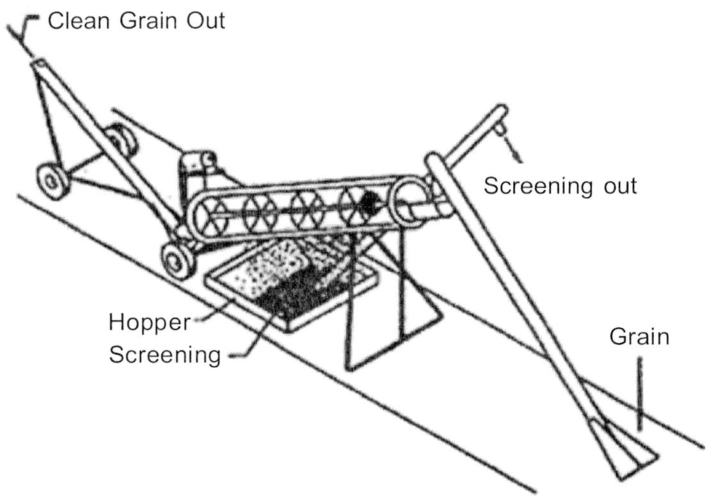

Fig. 1: Grain Cleaner

Capacity is influenced by grain moisture content, type of grain, slope of the cleaner screen, and the amount of fines and foreign material in the grain. Check with the cleaner manufacturer for specific operating guidelines and conditions.

Cleaning is easier and more complete at low flow rates and with dry grain. Cleaning can be accomplished after drying if the grain is dried and stored in separate bins. If the grain is not cleaned prior to drying, be prepared to spend more time cleaning and maintaining the dryer. This maintenance is very important to maintain the capacity of the dryer and to reduce the possibility of fires.

Screenings are good livestock feed. However, if the screenings are wet, they should be fed within two days. Also, take care when feeding fines from damaged grain. Moldy, soft, partially rotten grain particles are screened with fines, and may contain toxins harmful to livestock.

Grain Spreaders

After filling, the grain surface in the bin should be level. Peaked areas increase airflow resistance. Peaking also makes it more difficult and dangerous to monitor grain condition.

A properly adjusted and operated grain spreader can leave the top surface of the grain level with the fine material more evenly distributed throughout the grain mass. This results in more uniform airflow resistance throughout the bin.

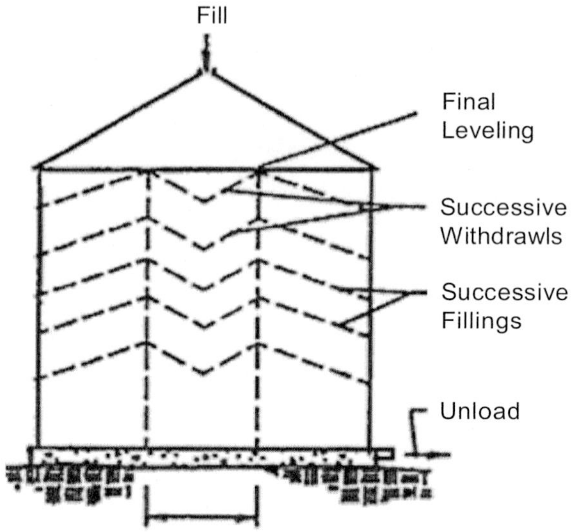

Fig. 2: Grain spreaders

One problem with using a grain spreader is that the grain tends to be packed tighter in the bin. Although this might seem good since more grain can be stored in each bin, packing increases airflow resistance, and the depth of grain may need to be decreased accordingly. A stirring device can loosen the grain, but excess use can contribute to grain damage. If a grain spreader is not used when filling a bin with grain that includes excess fine material, periodically unload a volume of grain from the center of the bin, a practice known as "coring." The objective is to create a 5 to 10 foot diameter depression in the center of each successive layer (Fig. 2.). The grain that is removed can be mixed with other grain and put back into the bin.

This practice does not remove the fines, but repeated unloadings help redistribute many of the fines that normally would accumulate in the center of the bin. This creates areas of relatively clean grain that improve airflow through the center of the grain mass. After the final filling, the grain surface should be leveled.

Moisture Content

If grain moisture content is too high, even the best aeration equipment and monitoring management will not keep the grain from spoiling - it only delays the inevitable. Recommended moisture contents for stored grain are listed in Table 1. These recommendations refer to the wettest grain in the bin, *not* the average moisture content.

For poor quality grain, the recommended maximum moisture levels should be 1 percent below that shown in the Table. Poor quality includes grain produced in hailed fields or droughty conditions and grain containing a high percentage of broken and cracked kernels due to improper combine adjustment.

Table 1: Maximum recommended storage moisture contents for aerated grain

Storage Period	Corn and Sorghum	Soybeans	Small Grain*
Fed by early spring	18%	13%	13%
Fed or marketed by June	15.5%	13%	13%
Stored up to one year	14%	12%	13%
Stored longer than one year	13%	11%	13%

*Since small grains are harvested in the summer and held through warm weather, moisture contents should be less than those for corn and sorghum.

If high temperature drying is used, operate the dryer at the lowest temperature possible. High temperatures reduce grain quality and make grain more susceptible to breakage and stress crack development. Excessive temperature also binds lysine in corn, making it unavailable to livestock.

High temperature drying should be discontinued when the grain moisture is about two percentage points above the desired final moisture content. The remaining moisture will be removed in the cooling process.

Cooling to ambient temperatures should take place over a four to six hour period. During this time the grain goes through a "steeping" process where moisture in the kernels redistributes and equalizes. Cooling the grain too rapidly can cause stress cracks; cooling too slowly encourages spoilage.

If grain is cooled in storage bin, do not turn the fan off until all the grain has been cooled, regardless of weather conditions. Holding warm grain in a bin for even a few days is a needless risk. Natural air drying is an excellent way to minimize grain damage. Natural air drying involves providing relatively high airflow rates, at least 1 cubic foot of air per minute per bushel of grain (cfm/bu), for extended periods (6 to 12 weeks), depending on the weather and initial grain condition. One critical aspect of natural air drying is that the initial drying front must be moved through the full depth of the grain bin as quickly as possible. To accomplish this, the fans must be operated continually. Drying costs and grain damage can be reduced if the grain is allowed to dry in the field. However, this must be weighed against the uncertainty of future weather conditions, and possible increased harvest losses.

Going into storage at the proper moisture content does not guarantee grain will remain at that moisture. Grain may be rewet as a result of bin roof or sidewall leaks. Moisture also can enter through downspouts from a bucket elevator or through hatches that have been left open.

Condensation also can cause localized increases in moisture content. Condensation, particularly on bin roofs and sidewalls, is common when warm grain (50°F or above) is cooled during cold weather (30°F or less), or when hot grain from a dryer is cooled in a storage bin. Minimize these condensation problems by cooling stored grain in stages during the fall; cooling hot grain from the dryer in a separate cooling bin before moving it to storage; and by providing adequate exhaust vents in the cooling bin.

Grain Temperature

Whether holding wet grain for a short period of time or storing dry grain for longer periods, it is important that grain temperature be controlled. Because both wet grain and molds respire and give off heat, aeration is needed to keep the grain cool and to slow mold growth. Aeration is the movement of relatively low volumes of air through stored grain. Airflow rates typically range from 0.1 to 0.75 cfm/bu. Properly aerated grain generally can be held safely about four times longer than non-aerated grain.

Aeration is needed to keep the grain mass at the desired temperature and to keep temperatures equalized, even if grain is dry and cool when placed in storage. Temperature differences within the grain mass create convection currents that can move and concentrate moisture in the top center of the bin. Problems caused by this moisture movement, or moisture migration, often become obvious in the spring when outside air temperatures begins to warm. The first indication of trouble is usually damp or tacky feeling kernels at the grain surface, followed by the formation of a crust.

Moisture migration is more of a problem in peaked grain because the moisture is concentrated in a smaller volume of grain. Moisture also moves by vapor diffusion from warmer to cooler areas in the bin. If grain is not properly cooled for winter storage, there is a tendency for moisture to move to the cool grain along the bin sidewall, causing spoilage. Moisture movement problems can be minimized by keeping grain mass temperatures equalized and within 10°F of the average monthly outside air temperature.

It is especially critical to cool grain from warmer harvest or summer storage temperatures. Cooling the grain early in the fall, and keeping grain temperatures below 60°F as long as possible into the summer, will help minimize insect activity and increase the chances of getting through the summer without having to fumigate the grain.

Grain Temperature and Moisture Migration

More grain is lost because of poor grain temperature management than for any other reason. Poor temperature management can subsequently affect moisture, which in turn increases problems with insects and molds.

Initial temperatures for grain storage vary because of seasonal harvest periods. For example, wheat harvested in midsummer is stored at 80 to 90°F while corn or sorghum harvested in late fall goes into storage with temperatures at 30 to 60°F. Grain typically has good insulation properties. As storage extends over several seasons, the center mass of grain remains about the same temperature as it was at the time of storage. As outdoor temperatures fluctuate with the season, the outer layer of grain changes in temperature. This causes slow movement of both air and moisture. Air and moisture movement in stored grain will occur when grain temperature and the outside ambient temperature differ. The direction and speed of this movement depends on the temperature gradient between grain and outdoor temperatures (Fig. 3).

This natural circulation occurs when heavy cool air drains down through the grain near the outer edge of the bin and moves upward after warming in the center. As air temperatures rise, air becomes lighter, its moisture holding capacity increases and it moves upward through the grain, collecting additional moisture. As the warmed air reaches cooled layers of grain, excess moisture condenses to increase grain moisture. This process is called moisture migration. Obviously, grain near the sunny side of the storage bin is warmer than that on the shady side, creating additional migratory currents.

Regardless of the time of year, the best rule is to maintain grain temperatures within 10°F of the outside ambient temperatures. Minimum grain temperatures should be 30 to 40°F. Crusting and potential insect and mold conditions can occur when the grain moisture level is at or below maximum storage recommendations at the time the storage unit is filled.

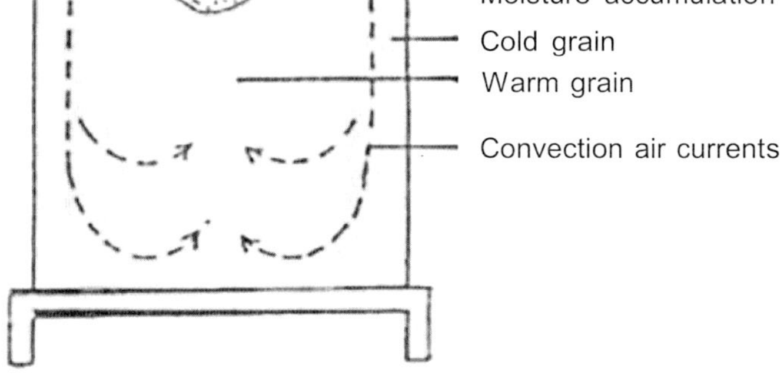

Fig. 3: Winter grain conditions and air-moisture migration.

Frozen Grain

The practice of cooling stored grain below freezing is not encouraged. There is no advantage to freeze stored grain. Disadvantages include condensation and aeration problems when the grain warms in the spring. During spring aeration, condensate near frozen chunks of grain may refreeze and block aeration in that zone. This blockage causes air channeling around the high moisture zone promoting mold formation.

Clean Grain

Only clean dry grain should be stored. Successful aeration for temperature and moisture control depends upon consistent, even air flow throughout the storage unit. Cracked grain, foreign material and fines create problems in stored grain. Broken or cracked kernels damaged during harvest are more subject to spoilage than sound kernels.

Airflow Rates

The required airflow rates through grain depend on the specific storage unit and management techniques used. Systems designed only for temperature control may require airflow rates as low as 1/20 cfm/bu (cubic feet of air per minute per bushel). High airflow rates require high capacity fans, use more energy and have high equipment and depreciation costs. The most common airflow rates are designed to provide between one-third and 1 cfm/bu, providing some drying capabilities in addition to rapid cooling.

Aeration rates for a particular bin depend on the bin type, air distribution system, desired grain moisture content and management practices, including close monitoring of changing conditions. Grain type and storage design affect static pressure, dictating fan size. For example, airflow resistance from 1/2 cfm/bu through 20 feet depth of corn results in static pressure of 2.6 inches. Fans moving air through wheat must overcome 5.9 inches of static pressure from the same bin size and airflow rate. A 25-foot depth of grain increases static pressure to 3.09 and 9.13 inches, respectively.

Fan Operation Time

Fan operation time required for each cooling or warming cycle depends on the aeration rates and the climate. To minimize storage problems, grain temperatures are kept within the bin as uniform as possible. When outside air temperatures differ by 10° or more from bin temperatures, aeration is done to reduce moisture migration.

Proper Ventilation

When using either positive pressure or negative pressure systems, it is necessary to provide adequate venting. Adequate venting provides the means for excessive moisture to escape and prevent structural damage. Closed vents and doors on suction systems during fan operation most likely will cause the roof to collapse. Likewise, in a positive flow system, fan operation in a closed bin will probably cause bulging and related structural damage.

Aeration Time

The actual time required may vary because of uneven or reduced airflow caused by climatic conditions or grain quality and can be determined only by monitoring the cooling or warming zone.

Variations in temperature change can be detected, indicating when the cooling or warming cycle is complete, by placing an adequate thermometer 6 to 12 inches into the grain in various locations.

Aeration systems should not be confused with the higher airflow system designed for natural air drying. Stored grain is aerated primarily to control grain temperature. Relatively low airflow rates, typically between 1/10 and 1/5 cubic feet of air per minute per bushel (cfm/bu), are adequate for this purpose. These airflow rates are not high enough to do a significant amount of drying. It is important that grain be stored at the proper moisture content and that the aeration system is managed to maintain favorable storage conditions.

Storable Grain Moisture Contents

The moisture content at which grain should be stored to maintain quality depends on the type of grain, the length of the storage period, and the temperature of the grain while in storage. Well-designed and managed aeration systems can be used to either temporarily hold higher moisture grain at cool temperatures or to store dry grain for several years.

Recommended stored grain moisture contents for corn, soybeans and sorghum are listed in Table 2. These recommendations assume the grain is aerated to prevent heating. Reduce the recommended moisture contents by 1 percent when storing low quality grain. This includes immature grain, severely cracked and damaged grain, and grain subject to previous insect or mold activity.

Table 2: Maximum recommended stored grain moisture contents for aerated grain

Storage Period	Corn and Soybeans	Sorghum	Small
Grain			
Marketed by June	15.5%	13%	—
Up to one year	14%	12%	13%
Over one year	13%	11%	13%

The moisture recommendations in Table 2 refer to the wettest grain in the bin, not the average moisture content. Hot spot development often starts in pockets of high moisture grain which can occur as the result of improper drying, condensation, bin roof or sidewall leaks, or moisture migration. Aeration can help disperse these moisture pockets and equalize moisture contents within the bin.

Moisture Migration

A common problem associated with long-term storage of dry grain is moisture movement within the grain mass, or moisture migration. Moisture migration is a result of convective air currents caused by temperature differences within the grain mass. This problem is especially severe when grain has not been properly cooled before winter storage and air currents move moisture to the top center area of the bin as shown in Figure 4.

Normally, moisture migration problems do not become readily apparent before spring when air temperatures begin to warm. The first indication of trouble is usually damp or tacky feeling kernels at the grain surface, followed by the formation of a crust. If this problem is discovered in time, the crust can be broken up and the aeration fan turned on to cool and dry the grain in the area of moisture accumulation. Better yet, the fan should be operated on a timely basis throughout the year to maintain uniform grain temperatures and prevent moisture migration.

Moisture Changes During Aeration

Aeration systems are not designed to dry grain; however, some moisture is uniformly removed from grain when it is cooled. The amount of moisture removed depends on both the temperature of the warm grain and the temperature of the cooling air.

Cooling grain taken directly from a high speed dryer can remove as much as 0.25 points of moisture for every 10°F the grain is cooled. A lesser amount of moisture is removed when stored grain is cooled about 0.15 points of moisture for every 10°F of cooling. For example, moisture contents would be reduced approximately 0.5 percent if grain were cooled from 65°F to 35°F during the fall. This relatively small amount of moisture is removed rapidly; requiring only the time needed to move a cooling zone through the grain (Table 3).

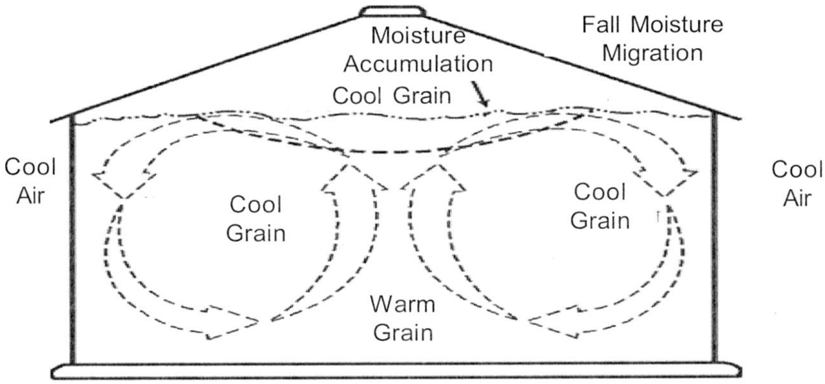

Fig. 4: Air convective currents that cause moisture migration problems in bins of improperly cooled grain.

A more commonly expressed concern is the amount of rewetting that can occur during an aeration cycle. This is normally not a problem if fan operation is limited to the relatively short period of time required to move a cooling or warming zone through the grain. In most cases, the effects of operating the fan during humid conditions are more than compensated for by the number of hours the fan operates under more favorable conditions.

Insect and Mold Control

Insects are generally not a problem in grain stored less than one year. However, if grain is to be stored for longer than a year, or if a bin has had an insect problem in the past, special precautions should be taken.

Table 3: Relationship between airflow rate and the approximate time required to movea cooling or warming zone through grain

System	TypeAirflow Rate, cfm/bu	Cooling or Warming Time, hours
Natural Air Drying	1	12-15
	3/4	16-20
Aeration	1/5	60-75
	1/10	120-150

Left untreated, an insect infestation eventually will lead to other storage problems. Insects give off moisture that can cause grain moisture contents to increase enough to create a mold problem. Mold activity will in turn raise temperatures and result in an increased rate of insect reproduction. Greater numbers of insects create more moisture, and the cycle is repeated at an ever-increasing rate.

Monitoring Grain Condition

Frequently monitor grain condition to verify that the desired temperature control is achieved. Regular examinations are essential if mold and insect activity are to be detected and controlled in a timely fashion. The method and frequency of checking will vary with time of year, initial grain condition, and aeration procedure. Generally, grain should be inspected at least once a month during the winter and every two weeks at other times of the year.

Extreme caution should be exercised when checking grain. Bridged grain is very dangerous, and collapse of the bridged area can result in suffocation. Always use a safety harness, lifeline, and grab rope, and have a second person outside the bin to assist in case of an emergency. Monitoring grain condition is extremely important during the summer because grain is held at higher temperatures and aeration conditions are less favorable. Grain temperatures should be checked and recorded at least every two weeks. Without records, it is difficult to tell whether elevated temperatures are caused by normally occurring outside temperatures or by heating due to mold activity. Use a grain probe to locate any moisture pockets where molds will develop rapidly as temperatures warm. Insect activity is also at a peak during the summer, and frequent checking is required if infestations are to be controlled before they develop into major problems.

Some areas and conditions to check when monitoring grain quality include:

- Grain surface for condensation, crusting, wet areas, molds, and insects.
- Bin roof for condensation and leaks.
- Grain mass for non-uniform temperatures, high moisture pockets or layers, molds, and insects.
- Exhaust air for any off-odors.

If problems are detected, they need to be evaluated and corrected as soon as possible. This may include cooling with aeration, further drying, or fumigation for insect control.

Maintaining proper grain temperature is important for both short-and-long term storage. Aeration is the most economical method of controlling grain temperature; however, aeration system capacities are seldom adequate to handle severe heating or moisture problems. Thus, a good aeration management program should be followed to both prevent storage conditions that favor mold and insect activity as well as react to problems that arise. Good managers further increase their chances for success by storing only clean, dry, high quality grain, and by complementing aeration management with a comprehensive grain-monitoring program. Thus it is critical to carefully manage stored grain to prevent deterioration and possible serious economic loss. This should include:

- A well-designed and properly operated storage system with adequate aeration capacity.
- Storing only clean grain at the proper moisture content and temperature.
- Checking the grain condition regularly and correcting problems before they get out of hand.

REFERENCES

Arbogast, R T and Mullen, M A (1990). Interaction of maize weevil and parasitoid (Anisopteromalus calandrae) (Hymenoptera: Pteromalidae) in a small bulk of stored corn. Journal of Economic Entomology, 83(6), 2463-2468.

Ashman, F. Webley, D J and Achillides, N Z (1974). Space treatments of a warehouse using dichlorvos. European Plant Protection Organisation Bulletin No.4(4), 429-445.

Baier, A H and Webster, B D (1992). Control of Acanthoscelides obtectus Say (Coleoptera: Bruchidae) in *Phaseolus vulgaris* (L.) seed stored on small farms. Journal of Stored Products Research, 28(2), 289-293 & 295-299.

Banks, H J. Annis, P C and Rigby G R (1990). Controlled atmosphere storage of grain: the known and the future. In: Proceedings of the Fifth International Working Conference on Stored-Product Protection, Bordeaux, France, September 9-14, 1990, pp 695-705.

Barak, A. V., and P. K. Harein. 1981. Losses associated with insect infestation of farm-stored shelled corn and wheat in Minnesota. Univ. Minn. Agr. Exp. Sta. Misc. Publ. 12, 94p.

Bell, C H (1976). The tolerance of developmental stages of four stored products moths to phosphine. Journal of Stored Products Research, 12:77-86.

Bisbrown, A J K (1992). Fumigable warehouses: design and construction. Natural Resources Institute Bulletin No.53. Natural Resources Institute, Chatham, UK; iv+28 pp.

Calderon, M, Donahaye, E, Navarro, S and Davis, R (1989). Wheat storage in a semidesert region. Tropical Science, 29: 9 1-11 0.

Champ, B R and Highley, E (1988). Bulk handling and storage of grain in the humid tropic. Proceedings of an international seminar, Kuala Lumpur, Malaysia, 6-9 October 1987. ACIAR Proceedings No.22, 296 pp.

Champ, B R. Highley, E and Banks, H J (1990). Fumigation and controlled atmosphere storage of grain. Proceedings of an international conference, Singapore, 14-18 February 1989. ACIAR Proceedings No.25, 301 pp.

Christensen, C M (1982). Storage of cereal grains and their products. American Association of Cereal Chemists Inc., St Paul, Minnesota, USA, 3rd edition, 544 pp.

Christensen, C. M., and R. A. Meronuck. 1988. Dry matter loss in yellow dent corn resulting from invasion by storage fungi. Plant Disease, 73:501-503.

Dermott, T and Evans, D E (1978). An evaluation of fluidized-bed heating as a means of disinfesting wheat. Journal of Stored Products Research, 14:1-12.

Desmarchelier, J M (1977). Selective treatments, including combinations of pyrethroid and organophosphorus insecticides, for control of stored product Coleoptera at two temperatures. Journal of Stored Product Research, 13(3):129-137.

Dobie, P (1984). In: Biological methods for integrated control of insects and mites in tropical stored products; ed. J A McFarlane. I. The use of resistant varieties. II. Integrated control: the role of biological methods. *Tropical Stored Products Information*, No.48: 4-8 and 37-45.

Donahaye, E (1990). Laboratory selection of resistance by the red flour beetle Tribolium castaneum (Herbst) to an atmosphere of low oxygen concentration. Phytoparasitica, 18(2):189-202.

Dunkel, F V (1992). The stored grain ecosystem: a global perspective. Journal of Stored Products Research, 28(2): 73-87.

Evans, D E (1987a). In: Integrated Pest Management, ed. A J Burner et al.; Academic Press: pp.425-461.

Evans, D E (1987b). The influence of rate of heating on the mortality of Rhyzopertha dominica (F.) (Coleoptera: Bostrichidae). Journal of Stored Products Research, 23:73-77.

Giles, P H (1965). Control of insects infesting stored sorghum in Northern Nigeria. Journal of Stored Products Research, 1:145-158.

Giles, P H and Ashman, F (1971). A study of preharvest infestation of maize by Sitophilus zeamais Motsch. (Coleoptera: Curculionidae) in the Kenya Highlands. Journal of Stored Products Research, 7:69-83.

Golob, P (1984). Improvements in maize storage for the small-holder farmer. Tropical Stored Products Information, 50:14-19.

Golob, P (1988). Current status of the Larger Grain Borer Prostephanus truncatus (Hoary) in Africa. Insect Science and its Application, 9(6),737-745.

Gough, M C and McFarlane, J A (1984). Aeration of stored grains: some psychrometric considerations. Tropical Stored Products Information, 50:32-35.

Grower, J H and Mullen, M A (1990). Effects of Xylocoris flavipes (HemipteM: Anthocoridae) releases on moth populations in experimental peanut storage. Journal of Entomological Science, 25(2):268-276.

Grower, J H and Press, J W (1990). Interaction of Bracon hebetor (Hymenoptera: Braconidae) and Trichogramma pretiosum (Hymenoptera: Trichogrammatidae) in suppressing stored product moth populations in small in-shell peanut storages. Journal of Economic Entomology, 83(3), 1096-1101.

Haines, C P (1984). In: Biological methods for integrated control of insects and mites in tropical stored products; ed. J A McFarlane. III. The use of predators and parasites. Tropical Stored Products Information, No.48:17-25

Haines, C P. Rees, D, Ryder, K, Sistyanto, S and Cahyana, Y (1990). Brown-rice bait bags for monitoring insect pest populations in bag stacks of milled rice as an aid to pest control decision-making. In: Proceedings of the Fifth International Working Conference on StoredProduct Protection, Bordeaux, France, September 9-14, 1990, pp 1351-1358.

Harris, K L and Lindblad, C J (Compilers) (1978). Postharvest Grain Loss Assessment Methods. American Association of Cereal Chemists in cooperation with The League for International Food Education, the Tropical Products Institute, the Food and Agriculture Organisation of the United Nations and the Group for Assistance on Systems relating to Grains After-harvest. 193 pp.

Hayward, L A W (1981) Structural features of warehouses adapted for long term storage in dry tropical climates. Tropical Stored Products Information, 40:14-23.

Hebblethwaite, M J (1985). The application of economics to pre- and postharvest pest control programmes in developing countries. Tropical Science, 25: 215-230.

Hodges, R J (1984). In: Biological methods for integrated control of insects and mites in tropical stored products; ed. J A McFarlane. II. The use of pheromones. IV. The use of insect diseases. V. The use of sterile insects. Tropical Stored Products Information, No.48, 9-15, 27-31 and 33-36.

Hodges, R J. Sidik, M, Halid, H and Conway, J A (1992). Cost efficiency of respraying store surfaces with insecticide to protect bagged milled rice from insect attack. Tropical Pest Management, 38(4), n-n + 6.

Hole, B P. Bell, C H. Mills, K A and Goodship, G (1976). The toxicity of phosphine to all stages of thirteen species of stored products beetles. Journal of Stored Products Research, 12: 235-244.

Holloway, G J (1986). The potency and effect of phytotoxins within yellow split peas (Pisum sativurn) and adzuki bean (Vigna angularis) on survival and reproductive potential of Sitophilus oryzae (L.) (Coleoptera: Curculionidae). Bulletin of Entomological Research, 76(2):287-295.

Joffe, A (1963). The effect of physical disturbance or 'turning' of stored maize on the development of insect infestation. South African Journal of Agricultural Science, 6(1), 5564.

Kamari, A and Yon, RM (1980). Temporary storage of wet paddy. pp 114-118 in: Grain Quality Improvement, Proceedings of the 3rd Annual Workshop on Grains Postharvest Technology, Kuala Lumpur, Malaysia, 29-31 January 1980.

Kossou, DK, Bosque-Perez, NA and Mareck, J H (1992). Effects of shelling maize on the oviposition and development of Sitophilus zeamais Motschulsky. Journal of Stored Products Research, 28(3):187-192.

McKenzie, B.A., L. Van Fossen and H.J. Stockdale. 1980. *Managing dry grain in storage.* Agricultural Engineers Digest, Iowa State University, Ames, IA 10 pp.

McFarlane, J A (1961). Malathion water-dispersible powder for the protection of bagged foodstuffs from insect infestation. Tropical Science, 3:114-125.

McFarlane, J A (1970). Treatment of large grain stores in Kenya with dichlorvos slowrelease strips for control of Cadra (Ephestia) cautella. Journal of Economic Entomology, 63(1), 288-292.

McFarlane, J A (1982). Damage to milled rice by psocids. Tropical Stored Products Information, 44:3-10.

McFarlane, J A (1988a). Pest management strategies for Prostephanus truncatus (Horn) (Coleoptera: Bostrichidae) as a pest of stored maize grain: present status and prospects. Tropical Pest Management, 34(2):121-132.

McFarlane, J A (1988b). Storage methods in relation to postharvest losses in cereals. *Insect* Science and its Application, 9(6), 747-754.

McFarlane, J A (1990). Differences between some strains of stored-grain beetles in their capacity to cause grain damage: possible implications for the management of pesticide resistance. Tropical Science, 30:357-371.

McGaughey, W H (1978). Moth control in stored grain: efficacy of Bacillus thuringiensis on corn and method of evaluation using small bins. Journal of Economic Entomology, 71(5), 835839.

Makanjuola, W A (1989). Evaluation of extracts of neem (Azadirachta indica A. Juss) for the control of some stored product pests. Journal of Stored Product Research, 25(4): 231-237.

Rees, (1990). The effect of temperature and relative humidity on population growth of three Liposcelis species (Psocoptera: Liposcelidae) infesting stored products in tropical countries. Bulletin of Entomological Research, 80:353-358.

Schulten, G G M (1976). Insects in stored maize ears. Abstracts on Tropical Agriculture, 2(6): 9-17.

Shires, S W (1982). Observations on Liposcelis bostrychophilus Badonnel (Psocoptera). Tropical Stored Products Information, 44:11-14.

Taylor, R W D (1991). Resistance to grain fumigants and future prospects for their use. Pesticide Outlook, 2(2): 22-24.

Taylor, R W D and Harris, A H (1989). Control of the egg stage of the Larger Grain Borer, Prostephanus truncatus (Horn) (Coleoptera: Bostrichidae), with phosphine. FAO Plant Protection Bulletin, 37(2): 67-70.

Taylor, R W D, Golob, P and Hodges, R J (1992). Current trends in the protection of stored cereals in the tropics by insecticides and fumigants. Proceedings of the Brighton Crop Protection Conference, Pests and Diseases Section, Brighton, UK, 281-290.

26

Storage of Pulses Under Modified/ Controlled Atmospheres

Gagandeep Singh and D.S. Sogi

Abstract

Pulses constitute an important component of the diet and are a major source of protein in vegetarian diet. India has largest area under pulse cultivation and produces a variety of pulses. A large portion of this produce gets spoiled due to insect and fungal attack. A variety of insecticides and fungicides are used for the control of pest during storage. There is an increasing trend towards the reduction in the use of pesticides in storage due to strong demand from consumers to choose food with no pesticide residue, development of insect resistance to pesticides and the impact of pesticides' use on environment. All these have posed a challenge to refine the existing management procedures of storage and commodity to find the optimal solution. Controlled/ Modified Atmosphere storage offers an alternative to the use of pesticides extends storage life and maintains product quality. The Controlled/Modified Atmosphere is created by altering the composition of the atmosphere in terms of CO_2, O_2, and N_2 concentration. High CO_2 produce undesirable conditions for insects and microbes and thus control their activity.

INTRODUCTION

Pulses are the edible fruits or seeds of pod bearing plants belonging to family of *Leguminosae* and are widely grown throughout the world. They have a high protein content ranging from 20-40 per cent. There is widespread protein-calorie malnutrition in developing countries and pulses can play an important role in bridging the protein gap. India depends greatly on pulses to meet its demand for protein.

Pulses and many other stored-products are living organisms in a state of dormancy that can remain unchanged for prolonged periods. However, dead grain rapidly loses its nutritive value, while non-dormant (germinating) grain cannot be stored. The grain remains viable and dormant because it is dry. High moisture permits the development of microorganisms that tend to kill the grain, while at very high moisture contents the grain begins to germinate. The use of fumigants and pesticides in the storage of pulses has lower levels of consumer acceptance and has also lowered pulse exports from India. So design of a storage structure should have all the advantages of enabling grain to be protected from pests without employing pesticides, which is the growing concern among the consumers.

Modified atmosphere (MA) technology is a standard technique for the post harvest handling of fruits and vegetables and could be extended to pulses to have a high quality produce available throughout the year. In less-developed countries modified atmosphere storage should provide an answer to the need for a less costly method of storage for food security of the rural populations. This can be achieved by supplying a storage solution at the farmer level, and thereby affording the farmer protection from seasonal fluctuations in grain prices. The basic advantage of modified storage in developing countries is its simplicity, obviating the need for dangerous insecticidal admixture procedures or fumigations both of which require high levels of expertise not usually possessed by the small-scale farmer.

Although this method has been well researched, its commercial use until recently has been rather limited. Possible reasons for its lack of commercial acceptance were its higher cost (in comparison with fumigation) and a lack of sufficient information on its reliability. However, cost comparisons between fumigation and alternative methods must be weighed against consumer acceptance, and as for its reliability, the extensive research on the effects of MAs on insects has provided a firm basis for the technology, and its potential to largely replace conventional fumigation. The application of MAs and fumigants is most appropriate for bulk storage of grain either on-farm, or at central storage installations that are gastight according to accepted standards. High value commodities requiring treatment in relatively small lots may be treated in specially designed fumigation chambers that may be adapted for MA application, or in flexible plastic containers which are also suitable for bag and crate storage. To complete this overview we shall describe the methods of MA generation and application, with emphasis on the main differences between them, and point out how our own work can promote the use of MAs as a fumigation alternative.

Modified atmosphere technology is a simple and environmentally safe way to manage post harvest insect and disease problems in cold storage. The use of low oxygen (anoxia) and high carbon dioxide and nitrogen environments simply smother insects. Anoxic environments could be a practical, cost-effective and safe way to kill all life stages of insects and mites. Factors that can influence the effectiveness of the modified atmosphere conditions include: atmospheric composition (percent of carbon dioxide, oxygen, and nitrogen); relative humidity (water loss); time; pest species; and pest life stage (Emeki *et.al.*, 2001)

The general effect of modified atmosphere conditions is to remove or limit the oxygen source of arthropod pests. Typically, as the oxygen concentration decreases, the rate of kills increases. Nitrogen gas alone induces hyperactivity followed by immobilization as a result of the anoxia or hypoxia. Carbon dioxide, like nitrogen, causes immobilization but it also induces opening of the spiracles. Thus, one of the primary effects of carbon dioxide is the loss of water by insects exposed to anoxic conditions, which ultimately leads to mortality. Carbon dioxide alone at concentrations of 35-60% gives almost 100% control of all life stages of most stored grain pests. Mites are generally harder to kill than insects. Age, condition (active or quiescent) and feeding status of arthropods can also influence their tolerance to elevated levels of carbon dioxide, or hypercarbic conditions. Hypercarbic conditions are also linked to physiological disorders such as decreased fecundity or longevity as well as disrupted metamorphosis. In addition to the effect of MA on pests, the effect on the crop plant is also important. Anoxic conditions have been used to disinfect cutflowers during storage. Most of the insecticidal MA research has been done at temperatures in the range of 0-12°C, while other work has been at temperature ranges of 20-40°C. The aim of this article is to focus the application of modified atmosphere technology to pulses.

METHODS FOR APPLYING MODIFIED ATMOSPHERES

Choice of atmospheric gas composition

A simple and descriptive graphical presentation to illustrate the relationship between exposure period, O_2 and CO_2 concentration (at normal atmospheric pressure), and mortality of different insects life stages, was compiled from the literature by Annis (1987). In his review, he proposed provisional dosage regimes at grain temperatures of 20-29°C.

The objective is to attain a composition of atmospheric gases rich in CO_2, or in N_2, and low in O_2, or a combination of these gases within the storage enclosure or treatment chamber, for the time necessary to control the storage pests. At present, the most widely used source for production of such atmospheric gas compositions is tanker-delivered liquefied CO_2 or N_2. Availability and suitability of this means of gas supply must be questioned when the gases are transported over long distances from an industrial production area to the storage site. Therefore, alternative potential methods of generating MAs should also be considered.

Supply of gases from tankers

When the target MA gas composition is <1% O_2 or high CO_2 concentration, a commonly used method is to supply N_2 or CO_2 from pressurized tankers. The practical aspects of purging grain storages have been described by Guiffre and Segal (1984) for CO_2, and by Banks et al. (1980) for N_2 and CO_2. A significant portion of the cost of applying MAs generated from tankers is for transportation and on-site purging. Bulk liquid gas is transported in conventionally insulated road tankers. For large-scale application of N_2 or CO_2, vapourizers are essential. These vapourizers consist of a suitably designed receptacle with a heating medium (electricity, steam, diesel fuel or propane), a hot-water-jacketed super-heated coil, and forced or natural draught. A forced-draught-type vapourizer with electrical super heating was found to be convenient.

Exothermic gas generators

For on-site generation of MAs by combustion of hydrocarbon fuel to produce a low O_2 atmosphere containing some CO_2, commercial installations -termed exothermic gas generators or gas burners- are available. Such equipment was originally designed for MA storage of fresh fruits. Their MA composition is designed to allow the presence of approximately 2-3% O_2 and to remove CO_2 through scrubbers. Therefore, their use in the grain industry requires several adaptations, like tuning the equipment to obtain an O_2 level of<1%; utilization to full advantage of the CO_2 generated; and removal of excessive humidity from the atmosphere generated. Combustion of propane yields approx. 13% CO_2 and of butane approx. 15% CO_2. The MA generated is more toxic than a N_2 atmosphere deficient in O_2. This is due to the presence of CO_2 in the MA causing hypercarbia, which together with hypoxia are synergistic in their effect on insect mortality.

Equipment has been designed to operate with open flame burners, catalytic burners, and as internal combustion systems. Full-scale field trials using open flame burners (exothermic MA generators), and catalytic burners to provide a low O_2 gas mixture, have proved successful. Open flame burners are capable of producing high gas flow rates at low O_2 tension. Consequently, the generated MA can be applied directly to purge the treated enclosure. On the other hand, catalytic systems reduce the O_2 concentration in the atmosphere by a fixed fraction during passage through the catalyst, and therefore should be used preferably in a recirculation system. The development of a modified internal combustion engine for MA generation has also been reported. In spite of its advantages over the open flame and catalytic burners as an easily operated, transportable and independent system, information on field application of such combustion systems is lacking.

On-site N_2 generators

Commercial equipment, termed also "pressure-swing adsorption" systems, using the process of O_2 adsorption from compressed air passed through a molecular sieve bed, is available. For continuous operation a set of two adsorbers is provided, which operate sequentially for O_2 adsorption and regeneration. N_2 at a purity of 99.9% can be obtained through regulation of inlet airflow. This method of N_2 generation is a relatively new approach in MA generation technology. Equipment is now being manufactured that is rated to supply an outlet flow rate of $120m^3/h$ at an outlet purity of 98% N_2. However, in view of the high capital cost investment involved, it would seem wise to undertake a long-term cost-benefit analysis to explore the justification of usage of these installations.

Biogeneration of MAs

Two principal forms of biogeneration of MAs exist, namely, "Hermetic storage" and "Assisted hermetic storage". The term "assisted hermetic storage" was introduced by Banks (1984) in order to define a process in which MA generation is assisted by a biogenerator source without sacrificing the commodity. The best known working example of assisted hermetic storage is that in use in China. With this method, removal of O_2 is achieved by recirculating storage gases through a closed system containing racks of moist grain and bran infected with a particular mould culture. This MA generation system merits further attention to explore potential applications at locations where a regular supply of industrial gases is non-existent or economically unfeasible. The basis for preparing these regimes was the time response of the most tolerant developmental stage of the most tolerant insect species. In the absence of *T. granarium*, a low O_2 regime should be based on the response of S. oryzae pupae, while the CO_2 regimes should be based on *Tribolium castaneum* adults and larvae (Annis, 1987; Navarro and Jay, 1987). More recently published information (Navarro and Jay, 1987; Reichmuth, 1987; Jay et al., 1990) indicates that further work is needed to enable precise dosage recommendations to be established for the application of MAs for the major stored-product insects under the wide range of intrinsic and extrinsic factors involved. Thus, recommended dosage regimes should be based on temperature ranges appropriate to specific climatic conditions and also to the dominant insect species found in the commodities involved. Aspects of commodity moisture content (Bell, 1987; Navarro, 1978), socio-economically acceptable control levels, the time-frame within which control must be accomplished, and the expected leak-rate standard in which the MA treatment will be performed will probably all play an important role in future recommendations.

CONCLUSION

The modified atmosphere storage has all the advantages of enabling grain to be protected from pests without employing pesticides. In the past, MA storage has been difficult to apply in practice because, the smaller the storage structure, the more critical is the importance of the hermetic seal, and the more expensive and difficult it is per unit of storage capacity to obtain the required gas tightness. This has often been the stumbling block, both for modification of traditional storage structures, and for the introduction of new ones to the village farmer. However, with recent advances in plastic manufacture, provide a sufficiently tight hermetic seal to preserve the pulses for prolonged periods.

REFERENCES

Annis P. C. (1987) Towards rational modified atmosphere dosage schedules: a review of current knowledge. Proc. 4th Int. Conf. Stored-Product Protection, Tel-Aviv, Israel, Sep. 1986, 128-148.

Banks H.J. (1984a) Current methods and potential system for production of modified atmospheres for grain storage. Modified Atmosphere and Fumigation in grain Storages, 523-542.

Banks H. J. (1984b) Assessment of sealant systems for treatment of concrete grain storage bins to permit their use with fumigants or modified atmospheres: laboratory and full-scale tests. CSIRO, Div. of Entomol., 38pp.

Banks, H.J., Annis, P.C., Henning, R.C. and Wilson, A.D. (1980) Experimental and commercial modified atmosphere treatments of stored grain in Australia, In: Shejbal, J. ed. Modified Atmosphere Storage of Grains. Amsterdam, Elsevier, 207-224.

Bell, C. H. (1987) Effect of grain moisture content on the establishment and maintenance of a low oxygen atmosphere containing carbon dioxide. BCPC Monograph No. 37. Stored Products Pest Control, 37:237-246.

Emekci, M., Navarro, S., Donahaye, E.J., Miriam Rindner and Azrieli A.(2001). Respiration of stored product pests in hermetic conditions. In: Donahaye, E.J., Navarro, S. and Leesch J.G. [Eds.] (2001) Proc. Int. Conf. Controlled Atmosphere and Fumigation in Stored Products, Fresno, CA. 29 Oct. - 3 Nov. 2000, Executive Printing Services, Clovis, CA, U.S.A. pp. 26-35

Guiffre, V., and Segal, A. I. (1984) Practical approaches to purging grain storages with carbon dioxide in Australia. In: Ripp, B.E., et al., ed., Modified atmosphere and fumigation in grain storages. Amsterdam, Elsevier, 343-358.

Jay, E.G., Banks, H. J., and Keever, D. W. (1990). Recent developments in modified atmosphere technology. In Champ, B. R., Highley, E., and Banks, H. J. (eds) Fumigation and Modified Atmosphere Storage of Grain. February 1989, Singapore. ACIAR Proc., 25:134-143.

Navarro, S. (1978). The effect of low oxygen tensions on three stored-product insect pests. *Phytoparasitica,* 6: 51-58.

Navarro S. and Jay.E, 1987, Application of Modified Atmosphere for controlling stored grains insects. In Proc. British Crop Protection Council No. 37, Stored Products Pest Control. Univ. of Reading. 229-236.

Reichmuth, C. (1987) Low oxygen content to control stored product insects. In: Donahaye, E., and Navarro, S., ed., Proceedings of the 4th International Working Conference on Stored-Product Protection. Tel Aviv, Israel, September 1986. 194-207.

27

Indigenous Grain Storage Practices Used By Rural Women

Geeta Thakur, Gayatri and C.S. Singh

Abstracts

About 60-70% of total food grain production is retained and stored by farmers for household consumption, use as seed, feed for cattle and payment of wages to labours etc. in our country. Yet the quality of grains stored is not so good, and hence, preserving the harvested crops without loss or damage has always been a problem of national importance. This study was conducted in Jabalpur district of Madhya Pradesh state. A tribal block namely Kundam was selected for study purpose. The responses of 120 women were recorded in pre-tested structural schedule through interview. Rural women are an important source of information on the use of indigenous practices for grain storage. The data reveals that the majority of respondents are using dry leaves of Neem, dry onion bulb, wheat straw, Nirgudi leaves, roots of bach etc. for storage of grain at household level. In our country, indigenous practices of grain storage have always been and are still followed, because it is supposed that these practices are useful in control of grain loss.

INTRODUCTION

The state, Madhya Pradesh is a rich source of various plants having numerous species of great importance due to wide variability in adophic and climatic conditions. The tribal uses plants for different purpose from time immemorial and more of their usages are still unknown to the people. The tribal farmwomen play a vital role in conservation of biodiversity and they have valuable knowledge of plants. They are an important source of information on the use of local plant species.

In our country indigenous practices of grain storage have always been and are still followed, because it is supposed that these practices are useful in control of grain loss and preserving the value of grain. Keeping in the view the above preposition, the study carried out with the objective of estimating the rural women's indigenous knowledge level regarding grain storage practices. The objective was to know the indigenous knowledge level of the rural women about the grain storage practices.

METHODOLOGY

This study was conducted in Jabalpur district of Madhya Pradesh state. Out of seven blocks of Jabalpur district, a tribal block namely Kundam was selected for study purpose on the basis of maximum area under forest. Out of the total the 197 villages, 12 villages & 10 women from each selected villages were selected by simple random sampling procedure after preparation of list of villages and list of women of each village who were having knowledge of using plants for storage of grains and seeds. Thus the responses of 120 women were recorded in pre-tested structured schedule through interview. The descriptive research design was employed and data were analysed by using frequency count and percentage.

RESULT & DISCUSSION

Information was collected on the use of indigenous practices of grain storage by the respondents. Data presented in Table-1 reveals that majority of the respondents (91.66%) using Neem leaves for grain storage. A healthy percentage of respondents (59.09%) mix dried onion bulbs in the grain during storage. Practices of keeping wheat straw on the storage sacs and gunny bags (49.16%) followed leaves of Nirgundi (33.33%).

Table 1: Use of indigenous practices for grain storage

S.No.	Common name of plant	Botanical Name	Practice	No. of respondents	Percentage
1.	Onion	Allium cepa	Dried onion bulbs in stored grain	65	59.09
2.	Wheat	*Triticum aestivum*	Wheat bhusa (straw) on the neck of bags	59	49.16
3.	Nirgundi	*Vitex nigundo*	Dry leaves in grain	40	33.33
4.	Bach	*Acorus calamus*	Dry roots of bach in rice	37	30.83
5.	Sitaphal	*Anona squamosa*	Seed of Sitaphal in grain	35	29.16
6.	Loung	*Syzygium auromaticum*	Piper in wheat	33	27.50
7.	Castor	*Ricinum communis*	Castor oil	29	24.16
8.	Turmeric	*Curcuma longa*	Dried turmeric slices	28	23.33
9.	Neem	*Azadirachta indica*	Dry leaves of Neem in grains	110	91.66

Roots of Bach (30.83%) especially for storing rice is present in rich seed of custard apple (29.16%), in grains and cloves (27.50%), castor oil for storage of seed by 24.16 % and dried slice of turmeric in grains by 23.33 % of respondents.

The study reveals that majority of the rural women had medium to high indigenous knowledge about grain storage practices.

REFERENCES

Jain, A.K. and Sharma, H.Q. (1994). Ethonobotanical study of Shahariya tribal tribal of Central India. Ethonobiology in human welfare: abstract of the Fourth International Congress of Ethonobiology, Lucknow, U.P., India, 17-21 Nov.pp81.

Tripathi, A.K. (1999). Depiction of ITK in Agricultural aspects prevailing in Vindhya region (M.P.). M.Sc. (Ag.) Thesis. Extension Eduation, JNKVV, Jabalpur.

SECTION - VII

GRADING AND TRENDS OF MILLING INDUSTRIES

28

Indian Grain Market - Effect of WTO

A.S. Bawa

Abstract

There is an urgent need to – create market driven linkages rather than do it alone by building new infrastructure. A lot needs to be done to improve infrastructure in terms of reducing the logistics cost, eliminating middlemen influence and eliminating the wastage of basic inputs. There is a need to blend technologies based on principle of dovetailing production for masses with the advantage of mass production, particularly in the area of grain processing, packaging and marketing which would help combine livelihood, security with agricultural competitiveness. Risk in crop production is not new to any age and the farmers have been bearing it since times immemorial. The farmers should be allowed to maximize their grain. High performance continuous and batch dryers remove moisture in a way that is predictable and reliable. The problems associated with moisture content may be harmful to international trade in grain and need to be addressed adequately by the standardization institutions.

INTRODUCTION

India is a super power in food security. It is very timely to conduct a review on such an important topic in light of the agreements such as WTO, GATT resulting in an emergence of a global market for different products requiring traditional commodity based production, handling and processing industry to adopt new practices, in order to ensure purity and quality of grains from seed to consumer.

Until recently our breeding programmes have been oriented towards higher yields to feed the masses and achieve self-sufficiency. Now we have moved from self-sufficiency to surplus, it is going to be an open market. In order to compete with the grains from competitive fields and price structure or else we will stand nowhere and even the countries like China will dump their grains in India, prices which our farmer will, in no way be able to compete. So, there is a need for a 2nd green revolution aiming at quality seeds / grain and much higher yield to be internationally competitive. Another aspect to be globally competitive would be to produce value added products from the existing product to meet the requirements of the importing countries. The products of a single variety may not be sufficient. Our strength is traditional Indian products with modified structure and taste profiles to suit the respective importer.

There is an urgent need to – create market driven linkages rather than do it alone by building new infrastructure. Build low asset based manufacturing models and innovative distribution systems. Customize the export orders – one cannot have a standard product for all markets. Low inwards to improve firm level competitiveness rather than looking at Government for tax reduction. A lot needs to be done to improve infrastructure in terms of reducing the logistics cost, eliminating middlemen influence and eliminating the wastage of basic inputs.

The world is watching India as destinations for investment in food technology not only because of its population but also because of the demand for traditional varieties of foods. It is the duty of food technologists to ensure that farmers get the best price. There is a need to blend technologies based on principle of dovetailing production for masses with the advantage of mass production, particularly in the area of grain processing, packaging and marketing which would help combine livelihood, security with agricultural competitiveness.

Risk in crop production is not new to any age and the farmers have been bearing it since times immemorial. The risk can be due to climatic uncertainties, international price movements, disease and pest infestation, nature of the products, seasonality in price and production, changes in government policies etc. To reduce this risk recently Union Agriculture Ministry has test launched Farm Income Insurance Programme (FIIP) with the main objectives:

1. Providing insurance coverage and financial support to the farmers in the event of failure of any of the notified crop as a result of natural calamities, pests and diseases.
2. Encouraging farmers to adopt farming practices, high value inputs and higher technology in agriculture.
3. Stabilizing farm incomes, particularly in disaster years.

This programme is expected to give relief to millions of marginal farmers, who are totally dependent upon farm income. It is expected to boost the grain production, as well as protect and nurture our agriculture too.

At present farmers can sell their produce in mandis notified by state Government. This system was introduced during times of deficient. Now it is leaching to farmers getting low prices as well as customers paying high prices. The profits go to agents controlling mandies. Industry is not permitted to purchase, produce from the farmer directly. Since the mobility of produce has increased, why should the farmer be forced to sell his produce in mandies / locally. They should be allowed to maximize their grain. Government procurement system can continue to exist for the benefit of small farmers while freedom can also be given to sell in the open market. It will also ease the burden on FCI and reduce handling and storage losses. Government has recently increased the price of wheat and rice to be sold to exporters because of increasing demands, which should not be the case to give free hand and boost to exports provided they are surplus.

Wheat and Maize are the major grains produced in the world at almost same levels 569 and 570 million tons, respectively followed by barley (50), Sorghum (63) and Rye (23). As far as wheat is concerned, China tops the world with 109 million tons followed by India 65 million tons and USA 64 million tons and European Union 94 million tons, Russia 34 million tons, Canada 25 million tons.

Trade exchange of wheat in terms of exports takes place only from countries like USA (30.1), Canada (18.2), European countries (14.6), Australia (13.8), Argentina (8.2) and others (8.9) million tons. While the importing countries are Egypt (6.8), Brazil, Japan, China, Algeria, Iraq, Indonesia, South Korea, European countries and Pakistan (2.7). Export of coarse grains is carried out by USA (54.6), Argentina (9.5), European countries (6.7), Canada (4.3), Australia (3.6), and others (14.8) million tons. While the importing countries are Japan (20.5), South Korea, Mexico, Chinese Taipei, Saudi Arabia, European countries, China, Egypt, USA and Malaysia (2.2) Million tons.

During the last few years shipments by products have been those of Wheat, Wheat flour, Rice, Maize, Sorghum and blended products to sub-Saharan Africa, Far East Asia, North Central America, North America, South America, North East Asia and Europe. While in terms of grains major shipment has been that of wheat (938 million tons) maize (66.5 million tons), barley (145 million tons), sorghum (6.6 million tons), oats (2.2) and Rye (16 million tons).

Recent assessment of International Grain Council has shown that world grain markets had become more volatile mainly due to demand uncertainties. Poor grains crops in Europe had more than offset high production levels in North America and Australia. Countries in Europe are taking special measures to restrain exports and facilitate imports to maintain domestic supplies of food and feed grains. Demands from some traditional wheat importing countries have been small because of their bumper crops.

STANDARDS

The term 'standard' refers to the measures that serve as a basis for making comparisons or judging the accuracy of unknown samples. Three types of standard are:

- Standard *specification* which define and specify a subject,
- Standard *test method* by which a specification is tested,
- *Grading* standard which allow a subject to be classified into more than one category.

Standards are established for a variety of purposes: a) for produce grading in agricultural marketing, or, increasingly, b) for the protection of consumers.

There are at least 330 specifications for cereals and cereal products at national and international level (over 50 countries or regions) of which at least 12 are applicable globally. Standard specifications provide criteria to characterize the nature of a commodity, usually on a pass or fail basis.

Most countries have a national standards institution which may issue specifications for commodities as well as methods of testing. Many countries adopt or modify international standards, e.g. International Organization for Standardization (ISO) standards, into their national system. Another source is the Codex Alimentarius Commission (Codex), which operates a committee to formulate standards on cereals, pulses and legumes.

Relative humidity and temperature of the maize with 15.0% m.c. loaded into a ship in northern latitudes at 4°C would have an inter-grain equilibrium relative humidity (ERM) of approximately 65% which is normally regarded as safe. However, if the grain is unloaded in a tropical country with an ambient temperature of 32°C, its inter-grain ERH becomes 75% - too high for safe storage. Heating and mould-damage took place when the grain was stacked at port of discharge, and confirmed that the standard m.c. of 15.5% at loading was too high for the conditions at the destination. A second shipment of grain dried specifically to approximately 14.5% m.c. suffered damage as well. In comparison, an intraAfrican importation with 12.2% mc did not suffer heating or deterioration on storage. Studies are continuing at NRI on the importance of integrating quality standards between suppliers and consumers.

The problems associated with moisture content may be harmful to international trade in grain and need to be addressed adequately by the standardization institutions. ISO 7979 goes some way towards this by acknowledging the principle of destination specific m.c. though without defining appropriate action. There are at least 420 standard test methods for cereals and cereal products at national and international level (over 50 countries or regions) of which at least 75 are applicable globally.

As with the specification of cereals, many countries modify or adopt international standards, e.g. International Organization for Standardization (ISO) standards, into their national system. Another important organization, particularly for the development of testing methods, is the International Association for Cereal Science and Technology (ICC). Other organizations issuing standard test methods include the Association of Official Analytical Chemists (AOAC) and the American Association of Cereal Chemists (AACC).

Lower moisture contents are required for certain destinations, in relation to the climate, and duration of transport and of storage. Grain that remains in the field for extended periods is at greater risk of weather damage. Early harvest supported by aeration and drying reduces the risk of harvest loss - particularly in a wet season.

Poor management of temperature and moisture in stored grain results in moulding, heating and insect activity. Aeration cooling, aeration maintenance and fumigation in sealed storages are key weapons in protecting grain against the enemies of grain quality.

Starting harvest early reduces the risk of field loss from weather damage and gives more management options. A strategy available every season is to start harvesting at marginal moisture levels (typically 2% to up to 3% high). This over-moisture grain is held under aeration then dried or blended to meet market requirements.

The benefits of an early harvest/aeration strategy include:

- Gaining extra harvest hours each day
- Returning to the field faster after rain
- Blending dry & over-moisture grain
- Holding wet grain longer before drying
- Evening-out grain moisture & temperature
- Preventing moulds & sprouting in grain
- Reducing quality loss from weather damage

COOLING

Cooling grain is only one beneficial function of aeration. The variety of equipment & purposes are:

- **Aeration-Maintenance** (selecting air to maintain grain quality)
- **Aeration-Cooling** (selecting air to cool grain quickly)
- **Aeration-Drying** (selecting air to remove moisture from grain)

The key difference between the 3 categories is the Quantity and Quality of aeration air needed to achieve a desired result. Aeration-drying typically uses 10 to 15 times the air-flow suitable for aeration-cooling. Aeration-cooling & maintenance systems reduce the grain bulk temperature to lower than average air temperature. Over time, they also create even conditions throughout the grain mass. The wide range of quality-related benefits includes:

- storing damp grain safely for longer periods
- avoiding 'hot spots', condensation & caking
- maintaining seed / malting barley viability
- protecting milling properties
- suppressing grain insect activity
- avoiding chemical residues
- inhibiting moulding & sprouting
- preserving colour and oil quality

Moisture damage in stored grain is often blamed on leaky silos. But a more common cause is moisture concentration as a result of uneven temperature profiles throughout the grain mass. 'Dry' grain also has plenty of moisture available to move around. Aeration cooling & maintenance is particularly effective in controlling moisture damage in stored grain.

High performance continuous and batch dryers remove moisture in a way that is predictable and reliable. These dryers need high air-flows. Purpose-designed aeration-drying systems provide opportunities to remove moisture from grain bulks at lower capital & operating cost than high-speed dryers. They add significant flexibility to a harvest system and are predictable in terms of what can be achieved under certain grain and weather conditions.

Moisture removal rates are relatively slow (days to weeks) and depend on availability of suitable drying air for extended periods. The system is most reliable when managing small batches of marginal-moisture grain as part of an early-harvest strategy. During wet harvests (when drying is needed most) aeration-drying delivers lower performance. Some growers use aeration as their primary insect management method. This is possible because aeration modifies temperature, moisture and air movement in a grain mass.

The result is an environment that deters insects from entering an aerated grain mass and dramatically reduces their activity if they are present. The effect of temperature on stored grain insects can be generalized as follows:

- Above 35° High temperatures limit growth
- 25°-35° Optimum insect population growth
- 15°-25° Activity reduces as temperature drops
- 10°-15° Population growth very slow
- 0°-10° No population growth/eventual death
- Below 0° Faster kill of stored grain insects

(These temperature effects are a general guide only and will vary between insect species.)

Aeration does not guarantee that there will be no live insects in stored grain. An integrated approach using aeration for grain quality management and sealed storage for highly effective insect control will deliver low risk for most operators. Sealed Silos offer reliable and low cost insect pest control. They also offer the added benefit of being able to treat grain without the risk of leaving unacceptable chemical residues. A correctly sealed silo maintains high levels of fumigant inside the silo for long periods of time. This ensures that all stages of the insect life cycle are killed resulting in insect-free grain.

Using fumigants in unsealed silos will not kill the early stages of the insect life cycle and there is a significant risk of encouraging insect resistance to fumigation. Most current practice involves use of Phosphine formulations (in tablet or blanket form) for fumigation. At present, the higher cost of Controlled Atmosphere (CA) systems using nitrogen or carbon dioxide limits their use to specialist applications including storage of organic grains.

CONCLUSION

Many market forces and technological changes will affect future grain storage practices on farms. Included amongst these are:

- Increased storage of grain on farms
- Wider use of aeration & drying in harvest management
- Continued demand for grain free of chemical residues
- Increased use of sealed storage & fumigants to control insects
- Tighter grain quality standards required by markets
- Increased application of aeration to protect grain quality
- Wider use of controllers to increase aeration efficiency

29

A Survey of Wheat Milling Units (Atta Chakki) in Delhi

Amar Singh, S.K. Jha, B.C. Srivastava, A. Pandeya and R.B. Gautam

Abstract

A survey of various wheat milling units (atta chakki) available around Indian Agricultural Research Institute, New Delhi was conducted to study the performance of these. The mills of various sizes (12", 16", 18", 20", 24" and 30" stone disc diameter) with horsepower ranging between 2 to 20 are in operation. Both horizontal and vertical disc types of mills are in use. The capacity of these units ranges between 10 kg/h to 200 kg/h. The stationary sieve, hand sieve, pedestal/exhaust fan are used for cleaning. The average wheat cleaning cost is Rs. 200 per ton. A small room of 8' x 10', 10' x 10' and 10' x 12' size is used for running the mills and storage of wheat grains. The grinding charges (Rs./q) of wheat, coarse grains (maize, sorghum, millet etc) and spices are 100, 200 and 1000, respectively. Sample of 500 g of wheat and flour were collected from each chakki and were analysed for wheat grain and flour quality. Infested, damaged grains were observed to be in 2-10% range. Sieve analysis of the flour was done and the average particle size, fineness modulus and uniformity index of the flour were calculated. Average particle size, fineness modulus and uniformity index were found in the range 0.1875 to 0.2023 mm, 1.052 to 1.297 and 0:1:9, respectively. Average particle size of the flour obtained from horizontal mills was found to be higher than vertical mills. Temperature of the flour at the outlet was observed in between 45 and 50°C at room temperature of 18.5°C.

INTRODUCTION

Wheat is the main staple food of India. It is processed into various products for human consumption in different types of agro processing industries. Wheat grinder commonly known as atta chakki is the most common machine in our country for flour making.

Quality in the case of foods is of special significance, which directly affects human health and safety. The process of quality control starts from raw material, process design, product composition, packaging, storage, distribution, consumer preparation and feedback. With increasing awareness, it is necessary for the Indian food processing sector to become quality conscious. For introduction of quality system in the industry, the government needs to frame the standards in the form of ISO 9000 and HACCP (Lelieveld et al., 2003). It, therefore, calls for sound technology and adequately trained personnel in handling food right from raw material through processing, packaging and transport to the consumption point.

Physical, chemical and microbiological are three main types of food contaminants. Food can become contaminated during various stages (Rasul, 2001). Finished products can get contaminated during storage and transport to shops, during the storage and preparation by the consumer.

The main route of contamination is through surface, air, water, people and pests. Processing, transport and packaging equipment may also contribute. The passage of food materials over a surface leaves residual food debris, which encourages the growth of microorganisms. Over the time these can multiply to sufficient numbers, which can affect the safety or quality of the food.

Chemical contamination may also results from the contact surfaces if they are not adequately rinsed after cleaning and disinfections. Lubricants, often unavoidable in equipment with moving parts, may contribute to chemical contamination. Non-product contact surfaces, such as floors, walls, ceilings, overhead beams and equipment supports are also important.

The present work was undertaken with the objective to survey the wheat grinding units (atta chakki) to collect information on quality of raw and finished products.

MATERIALS AND METHODS

A survey of ten wheat-grinding units located in Delhi was conducted to study the quality of wheat and flour. The sample of wheat and flour were collected and analysed. The quality of wheat was measured in terms of foreign matter content, moisture content, and hidden infestation percent and 1000 grain weight. The foreign matter and infested grains were manually separated. The quality of flour was related with the uniformity index, particle size and fineness modulus. The moisture content was determined by oven method (at 105±2°C for 24 h).

The information of milling units such as type, size, building size, sanitary condition, capacity, power required, cleaning and milling cost were collected. The temperature of flour was measured with a mercury thermometer. The flour samples were analysed for particle size, uniformity index and fineness modulus.

Sieve analysis was done using standard IS sieves No. 100, 60, 40, 30, 25, 18, and 9 with pan and cover. A sample of 250 g was shaked on sieves for 10 minutes. Percent material retained on each sieve was measured. The fineness modulus (FM) was calculated by dividing the sum of the weight fraction retained above each sieve by 100. A sample calculation of fineness modulus is given in Appendix I. Average size of grain (D) in mm was calculated by the following equation (Sahay and Singh, 1994): **$D = 0.135\ (1.366)^{FM}$**

The uniformity index (represented as X:Y:Z) is demonstrated on the basis of tabulated sieve analysis. The sum of X, Y and Z must total 10. The ratio X:Y:Z indicates the relative proportions of coarse, medium and fine materials. A sample calculation of uniformity index is given in Appendix II.

RESULTS AND DISCUSSION

The information on milling units visited during the survey is given in Table 1. The results revealed that there are wide ranges of milling units are available. These are available in the capacity range of 10 kg per hour to 500 kg per hour. The power requirement varies from 1.49 kW to 18.65 kW. These units are running in small building of 8 ‘ × 12’, 10' × 12' and 12’ × 12’ sizes. The difficulties in proper upkeep of stock are faced due to less space available. The sanitary conditions in these mills are poor. Cleaning cost of wheat is 20 paise per kg and the milling charge is Re. 1 per kg. The relation between size and power versus capacity is shown in Figs. 1 to 4 for both vertical and horizontal milling units. There is a linear relationship, which states that capacity increases with increases in disc size and power rating.

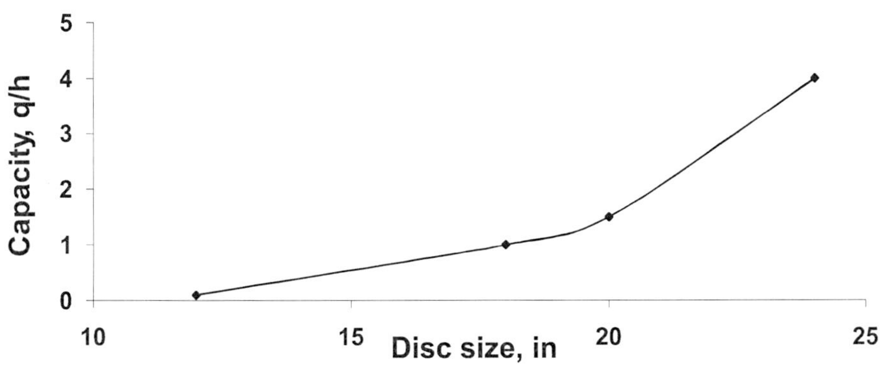

Fig. 1: Relationship between disc size and capacity of vertical mill

Grain quality

The wheat grain samples were analysed for moisture content, foreign matter content (organic and inorganic), number of insects, immature/ shrivelled / damaged/ discoloured grains, infested grains and 1000 grain weight. The results are given in Table 2. The results revealed that the wheat grains being milled are not pure. It contains immature grains, weed seed, damaged grains, infested grains and live adult insects. This results in poor quality of wheat flour.

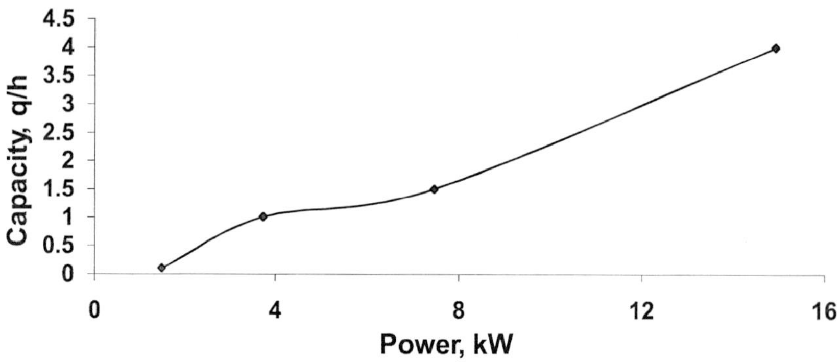

Fig. 2: Relationship between power and capacity of vertical mill

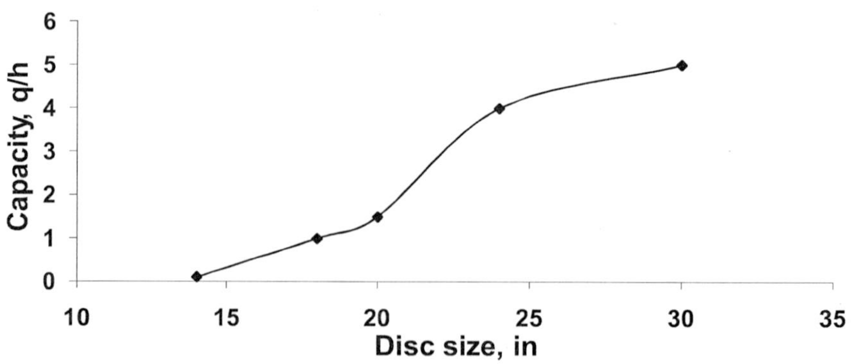

Fig. 3: Relationship between disc size and Capacity of horizontal mill

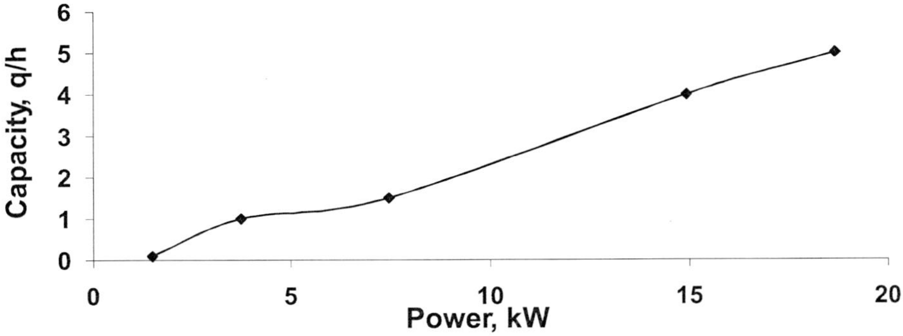

Fig. 4: Relationship between power and Capacity of horizontal mill

Table 1: Details of various milling units

Mill type	Disc Size, inch	Capacity, quintal/h	Electricity, kW	Price with electric motor, Rs.	Building size (floor area)
Horizontal	30	5	18.65	11500.00	12' x 12'
	24	4	14.92	8000.00	12' x 10'
	20	1.5	7.46	4200.00	10' x 10'
	18	1.0	3.73	4000.00	8' x 10'
	14	0.10	1.492	4300.00	8' x 10'
Vertical	24	4	14.92	17000.00	12' x 12'
	20	1.5	7.46	12000.00	12' x 10'
	18	1.0	3.73	7000.00	10' x 10'
	12	0.10	1.492	2800.00	10' x 10'

Table 2: Analysis of raw wheat

Sample	Damaged grain, %	Weed seed, %	Dehusked grain, %	No. of adult insect (beetle)	Moisture content, %	Weight of 1000 grains, g
A	9.40 (0.30)	0.46 (0.15)	7.1 (0.36)	6 (1.7)	10.8	34.5 (0.30)
B	1.66 (0.15)	0.53 (0.15)	0.3 (0.10)	0 (0)	11.2	45.9 (0.36)
C	1.93 (0.15)	0.20 (0.10)	0.7 (0.21)	1 (0)	10.6	43.4 (0.30)
D	3.56 (0.21)	0.43 (0.15)	1.4 (0.15)	4 (1)	10.3	40.2 (0.46)
E	8.33 (0.21)	0.36 (0.15)	2.4 (0.20)	5 (1)	11.0	37.8 (0.25)
F	3.46 (0.15)	0.30 (0.10)	1.7 (0.10)	8 (1)	10.7	41.4 (0.15)
G	7.26 (0.21)	0.26 (0.11)	2.4 (0.21)	4 (2)	10.1	38.9 (0.20)
H	4.50 (0.10)	0.50 (0.20)	1.5 (0.15)	6 (1)	11.0	40.0 (0.20)
I	3.73 (0.11)	0.60 (0.17)	1.3 (0.15)	1 (1)	11.2	39.3 (0.40)
J	2.96 (0.15)	0.37 (0.06)	3.7 (0.15)	2 (1)	11.3	42.2 (0.30)

(Values in parentheses indicate standard deviation based on three replications)

Flour Quality

The observation on flour quality defined in terms of particle size, uniformity index and fineness modulus are given in Table 3.

Table 3: Wheat flour quality obtained from different units

Sample	Mill type	Fineness modulus	Particle size, mm	Uniformity Index
1.	Vertical	1.258	0.1999	0:1:9
2.		1.248	0.1926	0:1:9
3.		1.297	0.1875	0:1:9
4.		1.198	0.1919	0:1:9
5.		1.139	0.1993	0:1:9
6.	Horizontal	1.139	0.1926	0:1:9
7.		1.052	0.2023	0:1:9
8.		1.115	0.1912	0:1:9
9.		1.128	0.1962	0:1:9
10.		1.231	0.1982	0:1:9

The fineness modulus gives an average size but does not indicate the distribution of fine, medium and coarse in the product and the average grain size is not proportional to the modulus index. The distribution of particles is given by uniformity index. The data presented in the Table 3 reveals that the milling units are converting wheat into very fine particles. The average size particle varies from 0.1875 to 0.2023 mm. The fineness modulus varies from 1.052 to 1.297, which indicates the fineness of the product. The uniformity index of the samples collected and analysed is 0:1:9, which indicate the distribution of coarse medium and fine. It is evident that product is having 9 part fine particle, 1 part medium and zero coarse particles.

CONCLUSION

These units should be located at proper places and proper building. The building should have good hygienic conditions. It should be constructed in such a way that microorganisms may not hide. The building should be such that it may be cleaned nicely. There should be separate room for storage of raw materials and cleaning. The dry cyclones, filters and wet scrubber should be installed so that the operator may be protected from inhaling fine particles from milling operations. The units should also be protected from environmental pollution.

Appendix I: Sample calculation of Fineness modulus and particle size

Weight of 1000 grains of wheat = 40.251

Sieve size	Opening size, mm	% Weight Retained	Weight Retained multiply by factor	Fineness modulus, FM	Average particle size, mm
100	1.000	00	7=00	125.86/100 = 1.2586	D=0132 $(1.366)^{1.2586}$= 0.1998 mm
60	0.592	0.032	6=0.192		
40	0.420	3.986	5=19.930		
30	0.296	2.508	4=10.032		
25	0.251	5.964	3=17.082		
18	0.177	17.59	2=35.180		
9	0.104	43.444	1=43.444		
0 (Pan)	00.00	26.476	0=00.00		
			125.86		

Appendix II: Sample calculation for uniformity Index

Sieve No. size, mm	Opening Retained	%, Weight by 10	Total divided to number	Rounded
100	1.000	00	4.408/10=0.4408	0
60	0.592	0.032		
40	0.420	3.986		
	Total	**4.408**		
30	0.296	2.508	8.472/10=0.8472	1
25	0.251	5.964		
	Total	**8.472**		
18	0.177	17.59	87.51/10=8.751	9
9	0.104	43.444		
(Pan)	00.00	26.476		
	Total	**87.51**		

REFERENCES

GOI. 2002. Department of Agriculture and Cooperation. (http://ww.kishanwatch.org/eng/statistics/march2002/indt_stat_area_prod_wht.htm)

Lelieveld, H. L. M., Mostert, M. A., Holah, J. and White, B. 2003. Hygiene in Food Processing. Woodhead Publishing Ltd., UK.

Rasul, N. 2001. Food Processing: Value addition and income distribution. Employment News, 10-16 November.

Sahay, K. M. and Singh, K. K. 1994. Unit Operations of Agricultural Processing. Vikas Publishing House Pvt. Ltd., New Delhi.

30

Grading of Grains - Its Evolution and Need

P.N. Maheshwari

Abstract

Grains in present context are food grains. The PFA Act includes cereals, millets, and pulses in food grains. At times, seeds of certain oilseeds like soybean are also referred to in the discussion of food grains. Grains have been cultivated for thousands of years. We have Grain Standards as established by the International Organization for Standardization (ISO), which are considered as International Standards. Another source is Codex Alimentarius Commission (Codex) under the aegis of FAO (UN), which operates a Committee to formulate standards that are referred to as Codex Standards. Many countries have a National Standards Institution, which develops National Standards for their main grain crops and other commodities and issues specifications along with methods of testing and analysis. Several countries, which are in developing status, adopt or modify international standards into their national system. The Grain Standards have been evolved and established for a variety of purposes but primarily to grade produce in agricultural marketing largely to facilitate the movement of grain, provide both sellers and buyers with guidelines to support financial transactions, and ensure that the produce will meet up with end-use requirements, or, increasingly to protect the consumers.

HISTORY

Grains have been cultivated for thousands of years. The fossilized pollen grains found in drill cores taken from lake sediments provide earliest evidence of maize being growing about 80,000 years ago in Mexico. Barley came in to cultivation about 10,000 years ago in the Fertile Crescent (which extended from Gulf of Arabia through Syria and Palestine to Egypt) of the Near East as barley remains have been found at Stone Age lake-dweller sites in Europe. Actual rice grains and husks have been excavated in India and in China that were 4500-5000 years old. Therefore, it is reasonable to believe that the agricultural revolution of the tenth millennium B.C. took place with the discovery of cultivable plants that could be grown near the dwellings and with domestication of animals to provide meat and milk. The first recorded farming communities were found in the region now in Iraq sometime in the ninth millennium B.C. These farmers could be considered as the founders of Grain Science and Technology. They grew wheat and other cereal grains besides keeping livestock and used irrigation systems to maintain their crops. Recent studies comparing ancient human remains and the foodstuffs that surrounded them for all parts of the globe suggest that farming communities began independently rather than starting and spreading from one community. The inhabitants of Northern India were growing wheat, rice and barley from about 2300 B.C.

It could be gathered from the Old Testament that in the grain market in Roman times, the problem of adulteration of wheat grain existed with seeds of a well - known weed Zizania, *Lolium temulentum*, which is a grass with its seeds similar to wheat. This weed seed infected with a fungus is toxic; about 200 gm of infected grain are enough to kill an adult. In smaller doses, it causes dizziness and mild delirium (temporary disorder of mental faculty, disturbance of consciousness). This prompted the Romans to make a regulation by declaring this weed seed a scheduled contaminant and imposing heavy penalties if a supplier's wheat contained any Zizania or *darnel* seeds in it. This is a very early example of Government setting standards of grain quality to protect the consumer.

The Romans cultivated wheat, barley, millets, and beans. Around 1200 A.D., the farming communities adopted the practice of dividing the collective harvest into three portions: one portion was stored in the house, another portion was reserved for the army, and rest was given to other people in accordance with their needs which, in essence, could be treated as a portion for the trade requiring cleaning and grading of grains. Maize, cultivated in Mexico and Central America by ~2500 B.C., was hybridized to increase the yields by ~1600 A.D. The agricultural system underwent changes in nineteenth century consequent to mechanization of various operations. There are indications that threshed wheat was piled up and thrown across the barn to separate heavy and light grains in the England in 1837. The inventions of the roller milling of wheat in 1870 and thereafter the milling of paddy in two distinct stages of shelling and polishing laid emphasis on thorough cleaning and proper grading of grains for which the systems were established largely in the twentieth century.

Present Status and Importance of Grain Standards

Today, we have Grain Standards as established by the International Organization for Standardization (ISO), which are considered as International Standards. Another source is Codex Alimentarius Commission (Codex) under the aegis of FAO (UN), which operates a Committee to formulate standards that are referred to as Codex Standards. Many countries have a National Standards Institution, which develops National Standards for their main grain crops and other commodities and issues specifications along with methods of testing and analysis. Several countries, which are in developing status, adopt or modify international standards into their national system.

The Grain Standards have been evolved and established for a variety of purposes but primarily to grade produce in agricultural marketing largely to facilitate the movement of grain, provide both sellers and buyers with guidelines to support financial transactions, and ensure that the produce will meet up with end-use requirements, or, increasingly to protect the consumers.

The establishment of quality and grading standards with efficient institutional framework for their effective implementation can be beneficial for producers/sellers and users/purchasers in the following ways:

- The standard will provide an unambiguous description of the quality/grade of the consignment in a uniform and meaningful manner, and assist in the formation of a legally - binding contract.
- Graded-grain is more equably priced than non-graded grain and brings stability to quality offered and market prices. Consumers, particularly processors, get assured quality for stable prices necessary for optimal performance.
- Grades and standards contribute to operational and pricing efficiency by providing buyers and sellers with a system of communicating price and product information, which leads to the development of effective and efficient marketing systems. The differences between different grades of a commodity, communicated to the markets, enable the buyers to signal which grades they are willing to purchase and at what premium or discount. Prices will vary for different grades depending upon the relative supply of and demand for each grade. The producers/sellers getting clear indication of the quality/grade required in the market can avail the opportunity for improvement of the quality/grade of produce to obtain potential rewards.
- The sanitary hazards associated with the inter-country movement of grain get reduced if clearly defined standards are enforced, particularly in relation to the prevention of spread of serious storage pests.
- The consumers' rights are protected through setting limits to the amount

of unsuitable or noxious material and thereby ensuring safety while protecting him/her from economic fraud via sub-standard material or product

Considerations in Standards for Grading of Grains

The subject matter of Grading of Grains/Grain Standards and Grades shall be considered, discussed and examined in the light of following:

1. Proximity of Consumer to Source of Grain

Over the years, the consumers have become demanding grains with particular qualities. Where consumers are close to the source of grain as in local trades/ markets, grading standards are rarely employed; their own preferences and the laws of supply and demand control the quality of grain which is assessed visually and is influenced by end-use, and the price is determined more by local rather than national factors. However, in transactions where large volumes of grains moving over long distances are involved as in regional or international trades, the buyer may never meet the seller or be able to examine the whole consignment. In such a situation, the consumer will have no direct influence over quality and therefore, the regulatory standards must be established and imposed to protect the consumers' rights. This is further relevant as some 12% of the world's grain production (~225 million tons) move across national borders highlighting the need for both national and international standards to ensure uniformity in quality and quantity of grain as well as freedom from environmental and health hazards.

As the quality of grains required for handling, storage and marketing forms a prerequisite part of the quality required for processing and utilization, the grading of grain is needed to be effected either at centres of aggregation/collection in assembly stage or when the grain move into storage rather than during warehousing or just before entering the market, as the grain should be sound, clean, and dry, free from infestation and impurities, for safe storage.

The grading of grain is not normally a separate marketing stage. It is a function to be provided by the storage / merchandising agency or the government. Since the government has to ensure effective implementation of the regulatory standards, it is appropriate to enact legislation assigning and empowering Commissioner of Agriculture to effect standardizing the grading of grains and to certify persons engaged in such activities, namely Grain Inspectors, as is effective in USA for the last two decades or so.

2. Grain Quality Characteristics

The criteria of grain quality must be established and accepted by all parties. The more important quality criteria relating to grading of grain are:

- Botanical, physical and chemical characteristics as specified / described in USDA guidebooks and given in Table 1.

Table 1: Grain grading parameters for wheat as per USDA

	Characteristics	
Botanical	Physical	Chemical
A. Species B. Varieties	A. Test weight B. Kernel weight (g/1000 kernels) C. Kernel size and shape D. Kernel hardness E. Vitreousness F. Colour (of the bran) G. Damaged Kernel H. Impurities / dockage I. Milling quality	A. Moisture content B. Protein content C. Protein quality D. Alpha amylase activity E. Fat acidity F. Crude fiber and ash

- Intrinsic varietal (colour, composition, bulk density, odour, aroma, size, shape) and environment- or process- induced (age, broken grain, chalky or immature grain, foreign matter, infested / infected grain, other grains, moisture content) parameters as employed in Codex and ISO specifications and reflected in standards for maize (Table 2) and wheat (Table 3), respectively.

Table 2: Codex Standard - Maize / Corn (Codex Standard 153-1998)

Criteria	Maximum tolerances (% m/m)
Abnormal or foreign odour	0
Moisture content	15.5
Blemished grain	7
of which : diseased grain	0.5
Broken kernels	6
Other grains	2
Foreign matter	2
Inorganic matter	0.5
Filth	0.1
Toxic or noxious seeds, heavy metals, microorganisms or poisonous or deleterious substances	Free from amounts hazardous to health

It is not unimportant to emphasize here that standards should be built on those characteristics that are important to both grain producers and end-users, and are easily recognizable and unambiguous, and can be accurately and uniformly measured and interpreted. The terminology used in the process must be simple and clear to understand.

Table 3: International Standard – Wheat (ISO 7970: 1989)

Criteria	Maximum tolerances (% m/m)
Foreign or deteriorated odour, additives, toxic substances	0
Pesticide residues, other contaminants	National limit or Co limit
Living insects	0
Moisture content	15.5*
Bulk density, minimum (kg/hl)	70
Damaged grain	15
of which: Broken grain	7
Shriveled grain	8
Unsound grain	1
Grain attacked by pests	2
Other cereals	3
Extraneous matter	2
Inorganic material	0.5
Harmful and/or toxic seeds, bunted grains and ergot	0.5
Ergot	0.05
Falling number, minimum	160

*Lower moisture contents are required for certain destinations, in relation to the climate, and duration of transport and of storage (Refer ISO 6322 for further information).

The grain characteristics such as varietal purity, soundness and vitreousness and grading factor such as foreign material (anything other than the very grain that remains in the sample after removal of dockage / waste material) are determined visually by rigorously trained personnel (Grain Inspectors); however, grading large samples of grain is a very tedious job and decisions are influenced by factors such as experience and expertise of the personnel, working conditions, fatigue etc. Consequently use of objective tools such as Machine Vision System (MVS) developed by Grain Scientists of the University of Manitoba is desirable in automation of the grain grading process. The MVS uses a video camera, which acts as an eye and the computer, which does the work of brain. It is claimed to be compatible with other automated on-line processing tasks, can work round the clock, can take dimensional measurements more accurately and consistently then human beings, can give an objective measure of variables such as colour, projected area, and shape which an Inspector can only assess subjectively and that too without contact to be hygienic and least damaging to delicate materials being inspected. A software to classify and grade grains is also now available to automate and mechanize the process of sample collection and image acquisition to assist the grain inspectors in grading grain.

There are at least 330 specifications at national and international level in over 50 countries and regions, of which at least 12 are applicable globally. These standard specifications are largely consumer-oriented to provide criteria to characterize the nature of a commodity, usually on a pass or fail basis, particularly in relation to wholesomeness or fitness of consumption. Some of these specifications constitute *grading* standards, which allow a grain to be classified into more than one category, in certain countries for certain purposes.

There are at least 420 standards test methods at national and international level in over 50 countries and regions, of which at least 75 are applicable globally. These test methods have been developed / are issued by International Organization for Standardization (ISO), International Grain Council (IGC), Association of Official Analytical Chemists (AOAC), American Association of Cereal Chemists (AACC), and National Standards' Institutions. The justification for such a large number of standard test method stems from the fact that there is a large diversity in grain characters and uses for grain. The precise test results of sample analyses have practical value only if representative sampling is done and test method is followed religiously. There are at least 30 standard methods dedicated to sampling, of which five or more are applicable internationally; the remainders are national or regional standards. The use of equipments and grain grading tools, developed to aid grain inspectors and specified for sampling and assessment of grain quality characteristics, objectively for analytical and subjectively for varietal and physical parameters, is essential to eliminate bias and legal hassles.

3. End-use Requirements

The grading standards for exports in regional / international trades differ in requirements than those justified for domestic market / end-use, particularly in the developing countries. The export specifications are generally more stringent in relation to physical conditions such as moisture contents as evident in ISO specifications for wheat (Table 3), standard of quality (test weight) and total foreign material including other cereal grains as evident in Canadian wheat specifications for export and primary grades of wheat (Table 4).

Table 4: Canadian specifications for Canada Western Red Spring wheat - export and primary grades*

Class Name	Grade	Minimum Test Weight (kg/hL)	Maximum Total Foreign Materials including other seeds
No. 1 CWRS	Export	79.0	0.40 % including 0.2 % other seeds
	Primary	75.0	0.60 % including 0.2 % other seeds
No. 2 CWRS	Export	77.5	0.75 % including 0.2 % other seeds
	Primary	72.0	1.20 % including 0.3 % other seeds
No. 3 CWRS	Export	76.5	1.25 % including 0.2 % other seeds
	Primary	69.0	2.40 % including 0.5 % other seeds

*Refer Official Grain Grading Guide of Canadian Grain Commission.

*Note:*Canadian regulations require export shipments to be *Commercially Clean* else permission of CGC is required for shipment of *Not Commercially Clean* consignments.

It could be argued whether a country should have a single standard for domestic, regional, and international markets or it should have separate standards for each one of these as end usage may vary from one country to the next. A single standard containing only a set of specifications may hinder either local market or export opportunities based on what the standard permits against the requirements. The use of different specifications for domestic and international markets requires greater detailing of the standard as well as system of assessing quality and price of grain consignments. However, there are no two opinions in my assessment for introduction of compulsory minimum quality standards or quality intervention standards for organized markets to prevent any mischief in trade and ensure consumer protection.

International trade of grain may be possible without standard, if the buyer and the seller are aware of each other's requirements. However, when country specific information and requirements are not known, a standard will provide guidelines that ensure the maintenance of quality and safeguard the consumers' rights. The need for detailed standard may be lessened if a country adopts the "Fair Average Quality" (FAQ) system of assessing quality and price of grain consignments. The FAQ system, as practiced by many national and international marketing agencies, for sale and purchase of a produce, is essentially subjective. Here, samples from different parts of the available stock of produce offered for sale (which may be scattered on farms or in warehouses throughout the producing area) are submitted under seal to independent assessors/public analysts for appraisal. After visual examination, the assessors select those samples which they consider representative of the bulk of the samples, mix them together and reduce the lot to a single reference sample, which is declared to represent the FAQ of the seller's stock. A portion of this sample may be used for certain objective tests but most of it is retained by an independent agency like Grains and Feeds Trade Association (GAFTA) for a specified period, during which transactions involving the produce should be completed. If there is any dispute over quality, the independent agency can be referred to for arbitration and the reference sample may be used as evidence. The results of FAQ assessment relate only to the crop, which has been sampled and only for the period agreed between the seller and buyer. The main advantage of FAQ is that it enables producers to dispose off most of their crop with minimum trouble and expense. Above all, standards should be sources of information intelligible to all and serving a clear function in the production and utilization of grain.

Indian Scenario

India has been a predominantly strong agricultural country. After attaining independence, it has been striving to develop its agricultural processing industry through successive 5-year plans for rapid economic growth and high employment generation. It is now recognized that India is world's second largest food producer accounting for over 600 million tons per year; the contribution of food grains being about 205 million tons with share of cereals of 185 million tons and legumes / pulses of 15 million tons. The Indian production of oilseeds is of the order of 25 million tons.

During the last six and a half decades of sustained efforts since independence, the Indian agro-industrial sector has evolved itself through several distinct and successive phases influenced by the economic growth, the changing consumer attitudes, and the unimaginative government policies. These phases can be grouped as under with predominant features as applicable in present context of grading of grains:

Phase	Year	Features
Basic Development	1947-1970	Qualitative and quantitative limitation of raw materials
Transformation	1970-1990	Promotion of Public Sector enterprises for food handling, storage, processing and marketing
Early Growth	1990-2002	Increase in pace of food export in light of liberalization and in tune with WTO regime
Consolidation	2002 and beyond	Commitment to harmonization of food standards and integrated regulatory regime, backward integration of processing with production sector, increased sophistication in consumer expectation for quality/ variety and consumer really becoming the market king

The consumer wants reliable quality, freedom from adulteration, and safety even in staple food grains, which are becoming driving forces for brand emergence and their acceptance in business. The Government of India began encouraging food grain exports from second half of 2000-01 in view of mounting stocks. Consequently, one can see a quantum jump in food grain exports in the last year from the following figures:

Year	Exports (million tons)	
	Wheat	Rice
2011-12	0.75	7.0
2012-13	1.5	8.0

The export of food grains is likely to get further impetus with development of bulk grain transportation and storage facility in India in about two years. At present, the food exports contribute about 17% *in value* of the country's exports, of which 19% is comprised of food grains, predominantly rice. The Government of

India has permitted contract farming of *Basmati* rice and also implemented *Price Advisory* for export of *Basmati* rice, which is expected to prevent the exporters from playing any mischief with *Basmati* shipments, particularly with regard to quality.

According to PFA Act, 1954 (as amended up to 2002), the food grains meant for human consumption are whole or broken kernels of cereals, millets and pulses.

- The food grains shall be free from *Argemone maxicana* and *Khesari* in any form.
- The food grains shall be free from added colouring matter.
- The food grains shall not contain any insecticide residue other than those specified in Table of *Rule 65* and the amount of residue in the food grains shall not exceed the limits specified in the said Table.
- The food grains meant for grinding / processing shall be clean, free from all impurities including foreign / extraneous matter.

Explanations

a. "Foreign matter" means any extraneous matter other than food grains comprising of

   i. Inorganic matter consisting of metallic pieces, sand, gravel, dirt, pebbles, stones, lumps of earth, clay and mud, animal filth and in the case of rice, kernels or pieces of kernels, if any, having mud sticking on the surface of the rice, and

   ii. Organic matter consisting of husk, straws, weed seeds and other inedible grains and also paddy in the case of rice

b. Poisonous, toxic and / or harmful seeds - means any seed which if present in quantities above permissible limit may have damaging or dangerous effect on health, organoleptic properties or technological performance such as dhatura (D. *fastuosa linn* and D. *stramonium linn*), corn cokle (*Agrostemma githago* L, *Machai Lallium remulenum linn*), Akra (*Vicia species*);

c. "Damaged grains" means kernels or pieces of kernels that are sprouted or internally damaged as a result of heat, microbe, moisture or weather, viz., ergot affected grain and kernel bunt grains;

d. "Weevilled grains" means kernels that are partially or wholly bored by insects injurious to grains but does not include germ eaten grains and egg spotted grains;

e. "Other edible grains" means any edible grains (including oil seeds) other than the one, which is under consideration.

In addition to the above, the following standards exist in the PFA Act:

		Wheat	Maize	Jawar and Bajra	Rice
Rule	A.18.06.	01	02	03	04
Definition	Dry, mature, sweet, clean & wholesome grains	*Triticum aestivum, T. vulgare, T. durum, T. sphaerococcum, T. dicoccum, T. compactum*	*Zea mays*	*Sorghum vulgare and Pennisetum typhyoideum,* respectively	*Oryza sativa*
Moisture	≤ (% by wt.)	14.0	16.0	16.0	16.0
Foreign Matter	(% by wt.)	1.0 of which not more than 0.25% by weight shall be mineral matter and not more than 0.1% by weight shall be impurities of animal origin			
Other Edible Grains	(% by wt.)	6.0	3.0	3.0	-
Damaged Grains	(% by wt.)	6.0 (including bunt and ergot affected grains not exceeding 3.0 and 0.5% by wt., respectively)	5.0	6.0 (including ergot affected grains not exceeding 0.05 % by wt.)	5.0
Weevilled Grains	(%)	10.0 (by count)	10.0 (by ct.)	6.0 (by wt.)	10.0 (by ct.)
Uric Acid	(mg / kg)	100	100	100	100
Aflatoxin	(mg / kg)	30	30	30	30
Vomitoxin/ Deoxinivalerol (DON)	Not more than 1000 µg/kg				
Provided that the total of foreign matter, other edible grains and damaged grains shall not exceed (% by wt.)		12.0	9.0	10.0	6.0

(Contd.)

		Masur (Lentil)	Urd (Black gram)	Moong (Green gram)	Chana (Bengal gram)	Other food grains
Rule	A.18.06	05	06	07	08	14
Definition	Sound, dry, sweet, seeds/grains clean & whole	*Lens culinaris /Ervum lens or L. esculenta*	*Phaseolus mungo*	*Phaseolus aureus, P. radiatus*	*Cicer arietinum*	Not specified in A.18.06
Moisture	(% by wt.)	14.0	14.0	14.0	16.0	16.0
Foreign Matter(% by wt.) be		1.0 of which not more than 0.25% by weight shall be mineral matter and not more than 0.1% by weight shall impurities of animal origin				
Other Edible Grains	(% by wt.)	3.0	4.0	4.0	4.0	6.0
Damaged Grains	(% by wt.)	5.0	5.0	5.0	5.0	5.0
Weevilled Grains	(% by count)	6.0	6.0	6.0	10.0	10.0
Uric Acid	(mg / kg)	100	100	100	100	100
Aflatoxin	(mg / Kg)	30	30	30	30	30
Provided that the total of foreign matter, other edible grains and damaged grains shall not exceed (% by wt.)		8.0	9.0	9.0	6.0	12.0

(Contd.)

Rule	A.18.06.	Arhar Dal 09	Moong Dal 10	Urd Dal 11	Chana Dal 12	Masur Dal 13
Definition	Sound, dry, sweet, clean and wholes ome split seeds/ grains of	*Cajanus cajan*	*Phaseolus aureus, P. radiatus*	*Phaseolus mungo*	*Cicer arietinum*	*Lens culinaris or Ervum lens or Ervum lens or*
Moisture	≤ (% by wt.)	14.0	14.0	14.0	16.0	14.0
Foreign Matter	(% by wt.)	1.0 of which not more than 0.25% by weight shall be and not more than 0.1% by weight shall be mineral matter impurities of animal origin				
Other Edible Grains	(% by wt.)	0.5	4.0	4.0	2.0	2.0
Damaged Grains	(% by wt.)	5.0	5.0	5.0	5.0	5.0
Weevilled Grains	(% by count)	3.0	3.0	3.0	3.0	3.0
Uric Acid	(mg / Kg)	100	100	100	100	100
Aflatoxin						
Provided that the total of foreign matter, other edible grains and damaged grains shall not exceed (% by wt.)		6.0	8.0	8.0	7.0	7.0

We do not have grading standards to classify a grain into more than one grades / categories.

It is also pertinent to mention here that focus should be on compliance of the regulatory standards rather than prosecution for non-compliance as is the case in PFA Act at present. We have multiplicities of laws with anomalies and even outdated provisions which should be done away with at the earliest. The suggestion of FICCI for highly focused enforcement of food laws for food safety and quality standards, based on other country models, deserves consideration by the Government, as we continue to face the problem of pesticide residues and heavy metal contaminants in our food supply.

"We have grossly under-developed quality assessment facilities with enforcement agencies, inappropriate personnel manning the inspection and quality monitoring system, undue delays in legal proceedings involving quality arbitration, procedure delays in clearing import and export shipments, delays in harmonizing the food standards and restrictions with those evolved by WHO-FAO Alimentarious Commission, and plethora of rules and regulations administered by multiple government agencies. There is also an increasing trend of blacklisting of Indian exporters and high percentage of rejection of Indian products by some major buyers, which is causing irreparable damage to the image of the country". These situations, pointed out by Dr. V.H. Potty in his article, "Indian Food Industry: Will the Sun Ever Rise", appearing in Indian Food Industry, Vol. 22(2), March-April, 2003 deserve prompt attention of one and all.

Index

A

B

C

D

H

I

K

L

M

N

O

P

Q

R

S

T

U

V

W

X

Z